Engine
Technology 1

THE BUTTERWORTH GROUP

UNITED KINGDOM	Butterworth & Co (Publishers) Ltd London: 88 Kingsway, WC2B 6AB
AUSTRALIA	Butterworths Pty Ltd Sydney: 586 Pacific Highway, Chatswood, NSW 2067 Also at Melbourne, Brisbane, Adelaide and Perth
CANADA	Butterworth & Co (Canada) Ltd Toronto: 2265 Midland Avenue, Scarborough, Ontario M1P 4S1
NEW ZEALAND	Butterworths of New Zealand Ltd Wellington: 31–35 Cumberland Place, CPO Box 472
SOUTH AFRICA	Butterworth & Co (South Africa) (Pty) Ltd Durban: 152–154 Gale Street
USA	Butterworth (Publishers) Inc Boston: 10 Tower Office Park, Woburn, Mass. 01801

First published 1981

© Butterworth & Co (Publishers) Ltd, 1981

British Library Cataloguing in Publication Data

Nunney, Malcolm James
    Engine technology 1.
    1. Internal combustion engines
    I. Title
    621.43   TJ785

    ISBN 0-408-00511-4

Typeset by Page Bros (Norwich) Ltd
Printed and Bound by
Trade Litho Book Printers Ltd, Bodmin, Cornwall.

Engine
Technology 1

M. J. Nunney CGIA, MSAE, MIMI

Preface

This textbook has been written primarily to cover the requirements of the Engine Technology 1 Unit (U77/406) of the Technician Education Council A8 programme. It may therefore be regarded as complementary to the already published *Vehicle Technology 1* by the same author.

The Unit subject matter has in certain respects been either updated or amplified, or both, so that *Engine Technology 1* should also satisfy the likely requirements of students outside the United Kingdom who are pursuing other courses in engine technology at a similar level. Again, the opportunity has been taken to offer a completely fresh and, it is hoped, interesting presentation of the subject that will encourage further study.

Although many of the line diagrams appearing in this book have originated from the author, grateful acknowledgement must be accorded not only to the Publishers for allowing the use of illustrations from certain of their other technical books, but also to the following organisations for supplying illustrations: Alfa Romeo (GB) Ltd., BL Austin Morris, Brown Brothers Ltd., Champion Sparking Plug Co. Ltd., Fiat Auto (UK) Ltd., L. Gardner & Sons Ltd., Honda (UK) Ltd., Laystall Engineering Co. Ltd., Lucas CAV Ltd., Mercedes-Benz (UK) Ltd., Mitsui Machinery Sales (UK) Ltd. (Yamaha), Nissan Motor Co. Ltd. (Datsun), Perkins Engines Ltd., Renault (UK) Ltd., Toyota (GB) Ltd., and Volkswagen (GB) Ltd. (Audi).

M.J.N.

Contents

1 Piston-type internal combustion engines

1.1 THE FOUR-STROKE PETROL ENGINE

The motor vehicle engine is basically a device for converting the internal energy stored in its fuel into mechanical energy. It is classified as an 'internal combustion' engine by virtue of this energy conversion taking place within the engine *cylinders*.

Since the term 'energy' implies the capacity to perform work, the engine is thus able to propel the vehicle along the road and, within limits, overcome unwanted opposition to its motion arising from rolling friction, gradient resistance and air drag. To facilitate this process the engine is combined with a 'transmission system', the functioning of which is discussed in Chapters 2 and 3 of *Vehicle Technology 1*.

The vast majority of car engines are of the reciprocating piston type and utilise spark ignition to initiate the combustion process in the cylinders. They also operate on the four-stroke principle in which the piston travels one complete stroke for each of the successive events of induction, compression, combustion and exhaust.

1.1.1 Historical background

An eponym is one who gives his name to something, and among the better-known eponyms in engine technology are Otto cycle,

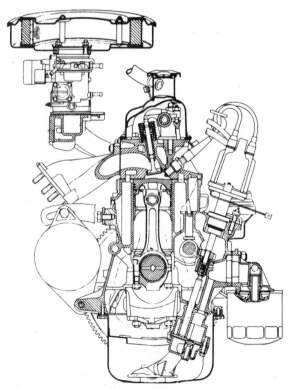

Figure 1.1 Cross-section of a four-stroke petrol engine (Fiat)

Diesel cycle and, perhaps to a lesser extent, Clerk cycle. Whether or not eponymic fame is entirely deserved in these cases has often been a matter of some controversy, it largely being a question of how successfully theoretical principles have been translated into working engines of practical form.

However, it would seem to be generally accepted that the first internal combustion engine to operate successfully on the four-stroke cycle was constructed in 1876 by Nicolaus August Otto (1829–1891). This self-taught German engineer was to become one of the most brilliant researchers of his time and also a partner in the firm of Deutz near Cologne, which for many years was the largest manufacturer of internal combustion engines in the world.

Although the Otto engine ran on gas, which was then regarded as a convenient and reliable fuel to use, it nevertheless incorporated the essential ideas that led to the development in 1889 of the first successful liquid-fuelled motor vehicle engine. This was the twin-cylinder Daimler engine, patented and built by the German automotive pioneer Gottlieb Daimler (1834–1900) who, like Otto, had been connected with the Deutz firm. The Daimler engine was subsequently adopted by several other car manufacturers and, in most respects, it can be regarded as the true forerunner of the modern four-stroke petrol engine.

1.1.2 The four-stroke petrol engine cycle

In this type of engine the following sequence of events is continuously repeated all the time it is running:

(1) The *induction* stroke, during which the combustible charge of air and fuel is taken into the combustion chamber and cylinder, as a result of the partial vacuum or depression created by the retreating piston.
(2) The *compression* stroke, which serves to raise both the pressure and temperature of the combustible charge as it is compressed into the lesser volume of the combustion chamber by the advancing piston.
(3) The *power* stroke, immediately preceding which the combustible charge is ignited by the sparking plug and during which the gases expand and perform useful work on the retreating piston.
(4) The *exhaust* stroke, during which the products of combustion are purged from the cylinder and combustion chamber by the advancing piston, and discharged into the exhaust system.

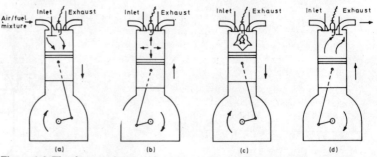

Figure 1.2 The four-stroke petrol engine cycle: (a) induction, (b) compression, (c) power, (d) exhaust

It thus follows that one complete cycle of operations occupies two complete revolutions of the engine crankshaft. Since energy is necessarily required to perform the initial induction and compression strokes of the engine piston before firing occurs, an electrical starter motor is used for preliminary 'cranking' of the engine. Once the engine is running the energy required for performing subsequent induction, compression and exhaust strokes is derived from the crankshaft and flywheel system, by virtue of its kinetic energy of rotation. 'Kinetic energy' is a term used to express the energy possessed by a body due to its mass and motion. The principle of an engine flywheel is therefore to act as a storage reservoir for rotational kinetic energy, so that it absorbs energy upon being speeded up, and delivers it when slowed down.

In the four-stroke cycle, the functions of admitting the combustible charge before its compression, and releasing the burnt gases after their expansion are performed by the engine inlet and exhaust valves. The opening and closing of the inlet and exhaust valves are not, in actual practice, timed to coincide exactly with the beginning and ending of the induction and exhaust strokes, nor is the spark timed to occur exactly at the beginning of the power stroke. At a later stage the reasons for these departures from the basic four-stroke operating cycle will be made clear.

1.2 THE FOUR-STROKE DIESEL ENGINE

It is now a matter of common knowledge that diesel engines have found widespread application in stationary power-generating units and as prime movers for marine propulsion, rail locomotion and, especially since the early nineteen-thirties, road transport. The essential difference between the petrol and the diesel engine is that the former relies on 'spark ignition' and the latter on 'compression ignition'. More specifically, the combustion process in the diesel engine is initiated by spontaneous ignition of the fuel when it is injected into a highly compressed charge of air, which has reached about 800°C. For this reason the diesel engine is often referred to as a 'compression ignition' engine, and, indeed, the terms 'diesel' and 'compression ignition' have long since become synonymous. Diesel engine combustion also tends to occur at 'constant pressure' rather that at 'constant volume', as in a petrol engine. This means that in the diesel engine the combustion pressure continues to rise steadily as the piston retreats and the cylinder volume increases, whereas in the petrol engine the combustion process is so rapid that there is very little movement of the piston while it occurs and hence very little increase in cylinder volume. Strictly speaking though, neither engine fits exactly into either of these categories.

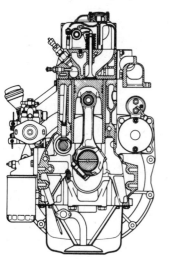

Figure 1.3 Cross-section of a four-stroke diesel engine (Hanomag)

1.2.1 Historical background

The diesel engine takes its name from Dr Rudolph Diesel (1858–1913), who was born in Paris of German parents. He became a student of mechanics and later entered the well-known engineering works of Sulzer Brothers in Winterthur, Switzerland. It was in the early eighteen-nineties that he developed his theories on what we now know as the diesel engine principle and subsequently took out various patents, including a British one granted in 1892. A few years later his theoretical work was embodied in a working engine of practical form built by the famous firm of MAN at Augsburg.

In fairness, however, it must be added that Diesel's concept of sparkless ignition was actually pre-dated by the pioneering work of an English engineer, Herbert Ackroyd Stuart (1864–1927). In 1890 he patented an engine operating on a similar principle, but which required a vaporiser surface at the end of the cylinder. For starting the engine, the vaporiser required the application of external heat. Hence, the first true compression ignition engine is generally attributed to Rudolph Diesel.

1.2.2 The four-stroke diesel engine cycle

In the four-stroke diesel cycle the following sequence of events is continuously repeated all the time the engine is running:

(1) The *induction* stroke, during which air only is taken into the combustion chamber and cylinder, as a result of the partial vacuum or depression created by the retreating piston.
(2) The *compression* stroke, in which the advancing piston compresses the air into the very small volume of the combustion chamber and raises its temperature high enough to ensure self-ignition of the fuel charge. This demands compression pressures considerably in excess of those employed in the petrol engine.
(3) The *power* stroke, immediately preceding which the fuel charge is injected into the combustion chamber and mixes with the very hot air, and during which the gases of combustion expand and perform useful work on the retreating piston.
(4) The *exhaust* stroke, during which the products of combustion are purged from the cylinder and combustion chamber by the advancing piston and discharged into the exhaust system.

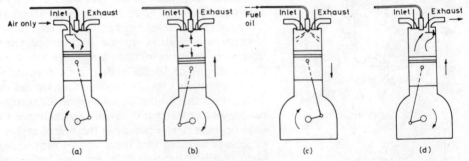

Figure 1.4 The four-stroke diesel engine cycle: (a) induction, (b) compression, (c) power, (d) exhaust

As in the case of the petrol engine, the timing for the opening and closing of the inlet and exhaust valves and also that for injecting the fuel, departs from the basic four-stroke operating cycle. Again, the reasons for this will receive consideration at a later stage.

1.2.3 Advantages and disadvantages
The following generalisations may be made relative to the use of diesel versus petrol engines in commercial vehicles and, more recently, cars:

(1) The diesel engine has better fuel economy than the petrol engine. This is because its 'thermal efficiency' is 30–36% as compared with the 22–25% of a petrol engine. The term 'thermal efficiency' represents the ratio of useful work performed by the engine to the internal energy it receives from its fuel.

(2) The diesel engine has generally proved to be more reliable, need less maintenance and also to have a longer life than an equivalent petrol engine. These advantages derive mainly from its sturdier construction and cooler running characteristics.

(3) Although a petrol engine develops its maximum power at higher rotational speeds than an equivalent diesel engine, the latter can provide better 'pulling power'. This is because the maximum turning effort or 'torque' exerted by the crankshaft of the diesel engine is greater and also better maintained over a wider range of engine speeds.

(4) A disadvantage of the diesel engine is that it tends to be heavy and bulky in relation to its power output. This is explained by the greater operating pressures and loads that have to be catered for in the construction of the diesel engine.

(5) The noise and vibration level of the diesel engine, especially under idling and low-speed operation, compares unfavourably with the petrol engine. Again, this is chiefly a function of the much higher cylinder pressures in the diesel engine. More recent diesel-engined motor cars have nevertheless been praised for their low level of noise at motorway cruising speeds.

(6) The diesel engine is sometimes criticised for having 'smokey' exhaust of unpleasant odour, although it is the invisible products of combustion in the exhaust gases of a petrol engine that are more harmful to the environment. Avoidance of a smokey exhaust with a diesel engine is largely a question of good driving technique, regular maintenance and proper adjustments.

(7) A safety consideration is that the fuel oil used in motor vehicle diesel engines is far less dangerously flammable than petrol, thus reducing the fire risk in the event of an accident. For taxation purposes, the fuel oil used in automotive diesel engines is referred to as DERV – an abbreviation of 'diesel engine road vehicle'.

(8) Finally, the basic cost of the diesel engine, together with its associated fuel injection equipment, is generally higher than that of an equivalent petrol engine.

1.3 THE TWO-STROKE PETROL ENGINE
As its designation implies, the two-stroke petrol engine completes its working cycle in only two strokes of the piston, so that a combustible charge is ignited at each revolution of the crankshaft. Although in its simplest construction the two-stroke

petrol engine needs no valves, the induction and exhaust process must be facilitated by a system of 'scavenging' or forcible clearing of the cylinder gases. This may either take the form of a separate engine-driven pump, or utilise the motion of the engine piston itself in a sealed crankcase. The flow of gases entering and leaving the cylinder is controlled by the reciprocating movement of the engine piston, which thus acts as a 'slide valve' in conjunction with 'ports' cut in the cylinder wall.

Familiar applications of the two-stroke principle to the petrol engine include motor-cycle engines, especially those used in smaller machines, marine outboard, horticultural equipment and small industrial stationary engines. Until recent years, this type of engine had also been favoured in Europe for the very inexpensive small cars, but it is now generally regarded as being obsolescent in car practice for reasons mentioned below.

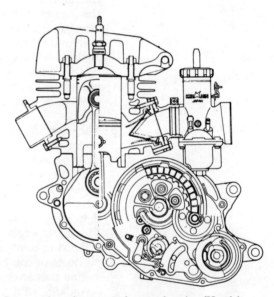

Figure 1.5 Cross-section of a two-stroke petrol engine (Honda)

1.3.1 Historical background

It is perhaps of interest to recall that the two-stroke and the four-stroke engine both originated in the late eighteen-seventies, so it might reasonably be assumed that both types of engine started out in life with an equal chance of success. The fact that the four-stroke engine became by far the more widely adopted type can probably be explained by its having a greater potential for further development. This is a criterion that can often be applied to rival ideas in all branches of engineering.

The first successful application of the two-stroke cycle of operation to an early gas engine is generally attributed to a Scottish mechanical engineer, Sir Dugald Clerk (1854–1932). It is for this reason, of course, that the two-stroke cycle is sometimes referred to as the 'Clerk cycle'. Dugald Clerk, like several other pioneer researchers of the internal combustion engine, was later to achieve high academic distinction, culminating in his election as Fellow of the Royal Society in 1908.

The Clerk engine was scavenged by a separate pumping cylinder and a few early motor vehicle two-stroke petrol engines followed the same principle, but it later became established practice to utilise the underside of the piston in conjunction with a sealed crankchamber to form the scavenge pump. This idea was patented in 1889 by Joseph Day & Son of Bath and represented the simplest type of two-cycle engine.

1.3.2 The two-stroke petrol engine cycle

In the two-stroke or Clerk cycle, as applied by Day, the following sequence of events is continuously repeated all the time the engine is running:

(1) The *induction–compression* stroke: A fresh charge of air and fuel is taken into the crankchamber as a result of the depression created below the piston as it advances toward the cylinder head. At the same time, final compression of the charge transferred earlier in the stroke from the crankchamber to the cylinder takes place above the advancing piston.

(2) The *power–exhaust* stroke: The combustible charge in the cylinder is ignited immediately preceding the power stroke, during which the gases expand and perform useful work on the retreating piston. At the same time, the previously-induced charge trapped beneath the retreating piston is partially compressed. Toward the end of the stroke the exhaust gases are evacuated from the cylinder, a process that is facilitated by the scavenging action of the new charge transferred from the crankcase.

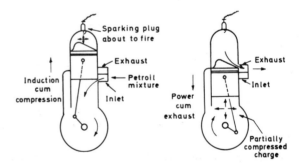

Figure 1.6 The two-stroke petrol engine cycle

The uncovering and covering of the cylinder ports by the piston, or 'port timing', is determined by considerations similar to those affecting the 'valve timing' of the four-stroke engine and will be explained at a later stage.

1.3.3 Advantages and disadvantages

The following generalisations may be made concerning the relative merits of two-stroke and four-stroke petrol engines in their basic form:

(1) The two-stroke engine performs twice as many power strokes per cylinder per revolution. In theory at least, this

might be expected to produce twice the performance of an equivalent-size four-stroke engine. Unfortunately, this is not realised in practice because of the difficulties encountered in effectively purging the exhaust gases from the cylinder and then filling it completely with a fresh combustible charge. The 'scavenging efficiency' of the basic two-stroke petrol engine is therefore poor.

(2) In performing twice as many power strokes per revolution, the two-stroke engine can deliver a smoother flow of power, but this may be less true at low engine speeds when irregular firing or 'four-stroking' can result from poor scavenging.

(3) An obvious practical advantage of the basic two-stroke engine is the mechanical simplicity conferred by its valveless construction, which contributes to a more compact and lighter engine that should be less expensive to make.

(4) Reduced maintenance requirements might reasonably be expected with the basic two-stroke engine by virtue of (3). There is, however, the well-known tendency for carbon formation to have a blocking effect on the exhaust ports, which impairs engine performance by reducing scavenging efficiency.

(5) The fuel consumption of the basic two-stroke engine is adversely affected by the poor cylinder scavenging, which allows part of the fresh charge of air and fuel to escape through the exhaust port before final compression of the charge takes place.

(6) There is a greater danger of overheating and piston seizure with a two-stroke engine, which can set a limit on the maximum usable performance. It is more difficult to cool satisfactorily, because it does not have the benefit of the second revolution in the four-stroke cycle when no heat is being generated.

(7) Lubrication of the two-stroke petrol engine is complicated by the need to introduce oil into the fuel supply to constitute what is generally termed a 'petroil' mixture. The working parts of the engine are thus lubricated in 'aerosol' fashion by oil mist in the air and fuel charge, and this tends to increase harmful exhaust emissions. It is for this reason that the two-stroke petrol engine is now obsolescent for cars.

1.3.4 Further developments

Frequent reference has been made to the inherently poor scavenging efficiency of the basic two-stroke petrol engine. The word 'basic' has been used deliberately and is intended to apply to the Day type of early two-stroke engine, which had a 'deflector-head' piston to promote a 'cross-scavenging' effect on the burnt charge leaving the cylinder. This not entirely successful scheme persisted until the mid-nineteen-twenties, when Dr E Schnuerle of Germany developed an alternative 'loop-scavenging' system. In this the deflector on the piston head is omitted and two transfer ports with angled passages are disposed on either side of, instead of opposite, the exhaust port. The loop-scavenge effect produced is such that before the two streams of fresh charge intermingle, they converge upon the cylinder wall at a point furthest away from the exhaust port, so there is less chance of escape.

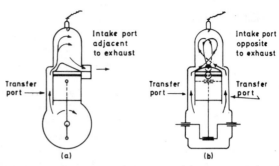

Figure 1.7 Early three-port scavenging systems: (a) cross-scavenging, (b) loop-scavenging

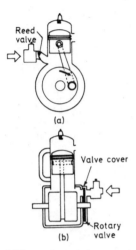

Figure 1.8 Later two-port scavenging systems: (a) reed valve (Yamaha), (b) rotary valve (Yamaha)

The Day type of early two-stroke engine also used what would now be classified as a 'three-port' system of scavenging. This system comprises inlet, transfer and exhaust ports, all in the cylinder wall, and necessarily imposes a restriction on the period during which a fresh charge of air and fuel may enter the crankcase.

To achieve more complete filling of the crankcase, the later 'two-port' system of scavenging is now generally employed. In this system only the transfer and exhaust ports are in the cylinder wall, the inlet port being situated in the crankcase itself and controlled by either an automatic flexible 'reed' valve, or an engine-driven rotary 'disc' valve. The two-port system of scavenging thus allows the fresh charge to continue entering the crankcase during the whole, instead of part, of the induction–compression stroke, albeit with a little extra mechanical complication.

1.4 THE TWO-STROKE DIESEL ENGINE

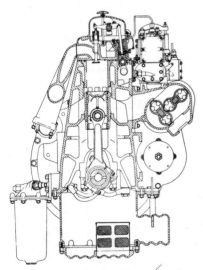

Figure 1.9 Cross-section of a two-stroke diesel engine (Foden)

The final type of engine that must be considered at this level of study is the two-stroke diesel. Since this type of engine may take several practical forms, our attention here will be concentrated on the high-speed version suitable for application to road transport.

Apart from the expected difference that air only is introduced into the engine cylinder prior to the injection of fuel oil, another departure from two-stroke petrol engine practice is that instead of using crankcase compression, a 'rotary blower' is used to charge the cylinder with low-pressure air. This type of blower was sometimes used in the past for 'supercharging' four-stroke-cycle engines. The distinction that must be made here, however, is that whereas a supercharger is used simply to increase the power output of a four-stroke engine, a similar blower is essential for a two-stroke diesel in order that it shall work at all.

Furthermore, a pair of exhaust valves are located in the cylinder head to provide a 'uniflow' system of scavenging. This means that there is no change in direction for the cylinder air stream, which is in contrast with the 'loop' system of scavenging described earlier for the two-stroke petrol engine. The two-stroke diesel engine is therefore mechanically more complicated.

1.4.1 Historical background

For about the first decade of its life the diesel engine was constructed solely in four-stroke form, following which fairly rapid progress was made in developing two-stroke versions for large, slow-speed, marine engines. Notable early examples were those built by the German firm of Krupp and the Swiss firm of Sulzer, the latter being first in the field in 1905. The two main reasons for this development were, first, the two-stroke diesel offered a reduction in weight and installation space, and second, the ease with which it was possible to reverse its direction of rotation – an important consideration for the marine engineer.

Although the two-stroke diesel engine was later developed for other marine, stationary power-generating and rail-locomotion purposes, its application to motor vehicles (at least in the United Kingdom) has been on a relatively limited scale. This can almost certainly be accounted for by its fuel consumption being generally greater than that of a corresponding four-stroke diesel engine, a matter of prime consideration in the road transport field. Nevertheless, there have been several high-speed two-stroke diesel engines successfully developed for automotive applications. Notable among these have been the uniflow-scavenged engines introduced by General Motors in America and Fodens Limited in the United Kingdom.

1.4.2 The two-stroke diesel engine cycle

In the basic two-stroke diesel cycle, the following sequence of events is continuously repeated all the time the engine is running and while the rotary blower is supplying air to the inlet ports of the cylinder.

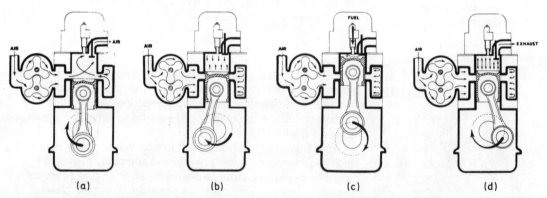

Figure 1.10 The two-stroke diesel engine cycle: (a)–(b) induction-compression, (c)–(d) power-exhaust (General Motors)

(1) The *induction–exhaust* event: Air only is admitted to the cylinder during the period the inlet ports are uncovered by the piston, which occurs towards the last quarter of the *power–exhaust* stroke and about the first quarter of the *induction–compression* stroke. During this part of the cycle, the exhaust valves are opened just before the ports are covered again. This sequence of exhaust valve events not only ensures that the exhaust gas pressure falls below that of the scavenging air supply, and thus prevents any return flow of exhaust gases, but also leaves the charge in the cylinder slightly pressurised prior to final compression. Hence, the

combination of uncovered inlet ports and open exhaust valves allows air to be blown through the cylinder, which removes the remaining exhaust gases and, by the same token, fills it with a fresh charge of air. Since neither the air nor the exhaust gases change direction in passing through the cylinder, the term 'uniflow-scavenging' can justifiably be applied.

(2) The *compression–power* event: The remaining three-quarter portions of the *induction–compression* and *power–exhaust* strokes occur in a very similar manner to that of the four-stroke cycle diesel engine; that is, the advancing piston compresses the air into the lesser volume of the combustion chamber and raises its temperature high enough to ensure self-ignition of the fuel charge. This is injected into the combustion chamber just before the piston begins to retreat on its *power–exhaust* stroke.

It should be noted that the operating cycle of the two-stroke diesel engine has been described in terms of 'events' rather than 'strokes' in order to assist understanding.

1.5 ENGINE NOMENCLATURE

To understand the information given in an engine 'specification table', such as those published in the motoring press, it is necessary to become familiar with some commonly-used terms.

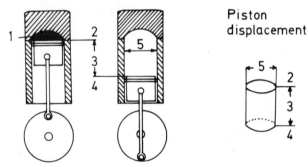

Figure 1.11 Engine nomenclature: (1) volume of combustion chamber, (2) top dead centre (TDC), (3) stroke, (4) bottom dead centre (BDC), (5) bore (Yamaha). Piston displacement: volume of gases displaced by the piston as it moves from BDC to TDC

1.5.1 Top dead centre

The term 'top dead centre' (TDC) is of general application in engineering and describes any position of a hinged linkage in which three successive joints lie in a straight line. In the case of a motor vehicle engine, top dead centre refers to the position of the crankshaft when the piston has reached its closest point to the cylinder head. This results in the main, big-end and small-end, bearings lying in a straight line. A motor vehicle service engineer often needs to establish top dead centre for checking the ignition and valve timing of an engine.

1.5.2 Bottom dead centre

The term 'bottom dead centre' (BDC) does, of course, refer to the opposite extreme of crankshaft rotation when the piston has reached its furthest point from the cylinder head.

1.5.3 Piston stroke

The term 'stroke' used in a general engineering sense refers to the movement of a reciprocating component from one end of its travel to the other. In the motor vehicle engine the piston stroke, therefore, is the distance travelled by the piston in its movement from TDC to BDC or, of course, vice versa, and is expressed in millimetres (mm).

1.5.4 Cylinder bore

In engineering practice the term 'bore' may refer to a hole through a bushing or pipe, or to the cutting of a large-diameter cylindrical hole, or to an actual measurement of the inside diameter of a hollow cylinder. It is the last-named with which we are concerned here, where the bore refers to the inside diameter of the engine cylinder expressed in millimetres (mm).

1.5.5 Piston displacement

This term refers to the volume of cylinder 'displaced' or 'swept' by a single stroke of the piston and is also referred to as 'swept volume'. It is expressed in cubic centimetres (cm^3) and may be simply calculated as follows:

$$V_h = \frac{\pi d^2 s}{4000} \, \text{cm}^3,$$

where V_h = piston displacement or swept volume (cm^3), d = cylinder bore (mm), s = piston stroke (mm).

1.5.6 Engine capacity

Here we are referring to the 'total' piston displacement or swept volume of all cylinders. For example, if the swept volume of one cylinder of an engine is say 375 cm^3 and the engine has four cylinders, then the 'engine capacity' is 1500 cm^3, or 1.5 litres (l). This can be simply stated as:

$$V_H = V_h z,$$

where V_H = engine capacity (cm^3), V_h = piston displacement (cm^3), z = number of cylinders.

1.5.7 Mean effective pressure

The term 'mean effective pressure' is used because the gas pressure in the cylinder varies from a maximum at the beginning of the power stroke to a minimum near its end. From this value must, of course, be subtracted the mean or average pressures that occur on the non-productive exhaust, induction and compression strokes. Engine mean effective pressure can be expressed in kilonewtons per square metre (kN/m^2).

1.5.8 Indicated and brake power

The most important factor about a motor vehicle engine is the rate at which it can do work or, in other words, the power it can develop. It is at this point that we must distinguish between the rate at which it might be expected to work (as calculated from the mean effective pressure in the cylinder, the piston displacement, the number of effective working strokes in a given time and the number of cylinders) and the rate at which it actually does work (as measured in practice when the engine is running against a braking device known as a 'dynamometer').

The significance of this is that the 'brake power' delivered at the crankshaft is always less than the 'indicated power', owing to internal friction losses in the engine. A simple expression for calculating in kilowatts (kW) the indicated power of an engine is as follows:

$$P = psAEz,$$

where P = indicated power (kW), p = mean effective pressure (kN/m^2), s = piston stroke (m), A = piston area (m^2), E = number of effective working strokes per second*, z = number of cylinders. (*Note: Since in the four-stroke cycle engine there is only one power stroke for every two complete revolutions of the crankshaft, the number of effective working strokes per second will correspond to one-half the number of engine revolutions per second.)

As mentioned earlier, a dynamometer is used in an engine testing laboratory to measure the brake power (or 'effective power') of an engine, because it acts as a brake to balance the torque or turning effort at the crankshaft through a range of speeds. A graph of the engine 'power curve' can then be drawn by plotting brake power values against engine speeds. Various standardised test procedures may be adopted in engine testing, such as those established by the American Society of Automotive Engineers (SAE), the German Deutsch Industrie Normale (DIN) and the Italian Commissione tecnica di Unificazione nell' Automobile (CUNA). In an engine specification table only the maximum brake power with corresponding crankshaft speed are usually quoted. For example, the high-performance engine used in the Mercedes–Benz 450SE/450SEL cars is claimed to develop, according to DIN, 165 kW at 5000 rev/min.

1.5.9 Engine torque

Also included on an engine performance graph is usually a 'torque curve', which is obtained by plotting crankshaft torque, or turning effort, against engine speed. The engine torque, of course, is derived from combustion pressure acting upon the cross-sectional area of the piston, the resulting force from which applies a turning effort to the crankshaft through the connecting rod and crankthrow arrangements. Engine torque, therefore, may be considered as the force of rotation acting about the crankshaft axis at any given instant in time.

Engine torque T may be expressed in newton-metres N m and generally reaches a peak value at some speed intermediate to

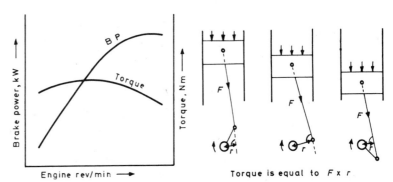

Figure 1.12 Engine power and torque curves

Figure 1.13 Engine torque, or turning effort

that at which maximum power is developed, the reason for this being explained at a later stage. An engine that provides good 'pulling power' is typically one in which maximum torque is developed at moderate engine speeds.

1.6 ENGINE COMPRESSION

Early internal combustion engines were very inefficient because they were provided with a combustible charge that was ignited at atmospheric pressure. However, it was recognised as early as 1838 by William Barnett in this country that compression of the charge before combustion was advantageous. Nearly 25 years later, a French railway engineer with the splendidly-sounding name of Alphonse Beau de Rochas was granted a patent in respect of several ideas that related to the practical and economical operation of the internal combustion engine. Among these ideas he stated a requirement for the maximum possible expansion of the cylinder gases during the power stroke, since the cooler they became the more of their energy is transformed into useful work on the retreating piston. To assist in this aim, there was a further requirement for the maximum possible pressure at the beginning of the expansion process, which the motor vehicle service engineer recognises as the 'compression pressure' of an engine. The first successful engine to utilise this principle was, as mentioned earlier, the Otto engine.

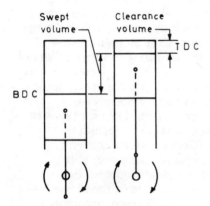

Figure 1.14 Engine compression ratio

Within certain limitations that will be better understood at a later stage, a 'high-compression' engine is relatively more efficient than a 'low-compression' one in terms of either improved fuel economy, or greater power. This is simply because it better utilises the internal energy received from its fuel or, in other words, it possesses a higher 'thermal efficiency'. It is this particular feature of engine operation that explains the importance of checking compression pressures before attempting to tune an engine in service.

1.6.1 Compression ratio

The extent to which the air and fuel charge in a petrol engine and the air alone in a diesel engine is compressed, prior to the power stroke, is known as the 'compression ratio' of an engine. It is calculated as the ratio of the total volume enclosed above the piston at BDC to the volume remaining above the piston at TDC.

The term 'swept volume' has already been explained in §1.5.5 and a complementary term is 'clearance volume', which is that volume remaining above the piston when it reaches TDC. In some modern engines the combustion chamber is formed mainly in the piston head, so that the clearance volume is concentrated within, rather than above, the piston. Hence, the compression ratio (usually abbreviated to CR) may also be expressed as the 'swept volume' plus 'clearance volume' divided by the 'clearance volume', as follows:

$$\varepsilon = \frac{V_h + V_c}{V_c},$$

where ε = compression ratio, V_h = cylinder swept volume (cm^3), V_c = combustion space clearance volume (cm^3).

The calculated compression ratios for petrol engines are typically in the range 8 to 9.5:1 and for diesel engines 16 to 22:1. It should be appreciated, however, that in the petrol engine the calculated compression ratio is realised in practice only when the engine is running with a wide open throttle, whereas the diesel engine nearly always runs unthrottled.

Furthermore, in the case of a two-stroke engine V_h is generally considered as the cylinder volume from the point of exhaust port closing to TDC, which is therefore less than the swept volume of an equivalent four-stroke engine and gives a lower compression ratio.

2 Structure and mechanism of the petrol engine

2.1 ENGINE COMPONENTS AND THEIR FUNCTIONS

It is convenient to introduce the various components of the petrol engine in the following groups:

(1) Cylinder block and crankcase
(2) Piston and rings
(3) Connecting rod and bearings
(4) Crankshaft assembly and bearings
(5) Cylinder head and gasket
(6) Valve train and timing drive
(7) Engine support mountings.

2.1.1 Cylinder block and crankcase

In combination, the cylinder block and crankcase form the main structural component of the engine and perform several important functions, as follows:

(1) Each cylinder must act not only as a pressure vessel in which the process of combustion can take place, but also as a guide and bearing surface for the piston sliding within it.
(2) Since the engine cylinders have to be cooled effectively, the cylinder block must also form a jacket to contain the liquid coolant.
(3) The crankcase provides an enclosure for the crankshaft and various other parts of the engine mechanism and must preserve accurate alignment of their supporting bearings under all operating conditions.
(4) Pressure conduits in the form of drilled ducts must also be incorporated in the crankcase to convey oil to the engine working parts.

It has long since been established practice to combine the cylinder block and crankcase into a single unit, this generally being termed 'monobloc' construction. The historical origins of this form of construction date back to the early nineteen-twenties, when there was a general trend towards simplification of the engine structure. Until then the cylinder block and crankcase were produced as separate units, in cast iron and aluminium alloy respectively, and then bolted together.

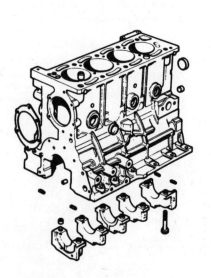

Figure 2.1 Cylinder block and crankcase (BL)

In relation to modern engine design, the monobloc construction provides a very necessary rigid foundation for the engine and reduces manufacturing costs. It should be added, however, that this particular form of construction is not always the best compromise for heavy-duty diesel engines, which will be considered at a later stage.

2.1.2 Piston and rings

The main function of the piston itself is twofold, as follows:
(1) It acts as a moving pressure transmitter by means of which the force of combustion is impressed upon the crankshaft,

through the medium of the connecting rod and its bearings.

(2) By supporting a gudgeon pin the piston provides a guiding function for the small end of the connecting rod.

The piston assumes a 'trunk' form to present a sliding bearing surface against the cylinder wall, which thus reacts the side thrust arising from the angular motion of the connecting rod. Since the piston is a major reciprocating part, it must of necessity be light in weight to minimise the inertia forces created by its changing motion – bearing in mind that the piston momentarily 'stops' at each end of its stroke! Another, perhaps obvious, requirement is that the piston must be able to withstand the heat of combustion and should operate quietly in its cylinder, both during warm-up and at the normal running temperature of the engine.

To perform its sealing function efficiently, the upper part of the piston is encircled by flexible metal sealing rings, known as the 'piston rings', of which there are typically three in number for petrol engines. In combination the piston rings perform several important functions, as follows.

(1) The upper 'compression' rings must maintain an effective seal against combustion gases leaking past the pistons into the crankcase.

(2) These rings also provide a means by which surplus heat is transmitted from the piston to the cylinder wall and thence to the cooling jacket.

(3) The lower 'oil control' ring serves to control and effectively distribute the lubricating oil thrown on to the cylinder walls, consistent with maintaining good lubrication and an acceptable oil consumption.

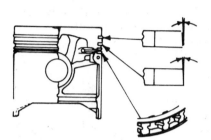

Figure 2.2 Piston and rings (Toyota)

2.1.3 Connecting rod and bearings

The function of the connecting rod and its bearings is to serve as a constraining link between the reciprocating piston and the rotating crankshaft. A fair analogy of this particular conversion of motion is to be found in the pedalling of a cycle, where the knee, calf and foot of the cyclist may be likened to the piston, connecting rod and crankpin of an engine.

The connecting rod is attached at what is termed its 'big-end' to the crankpin and at its 'small end' to the gudgeon pin of the piston. Each connecting rod big-end bearing is divided into two half-liners, so as to make possible its assembly around the crankpin. The big-end bearing housing is therefore formed partly by the lower end of the connecting rod and partly by a detachable cap, the two halves being bolted together. No such complication arises in the case of the small-end bearing arrangements, this end of the connecting rod being formed as a continuous eye.

As a consequence of its reciprocating and partly rotating motion, the connecting rod is subjected to appreciable inertia forces. Its detail design must therefore be such as to ensure the maximum rigidity with the minimum weight.

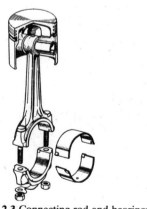

Figure 2.3 Connecting rod and bearings

2.1.4 Crankshaft assembly and bearings

The crankshaft represents the final link in the conversion of reciprocating motion at the piston to one of rotation at the flywheel. In the case of the multi-cylinder engine, the crankshaft has to control the relative motions of the pistons, whilst simultaneously receiving their power impulses.

A one-piece construction is most commonly used for the motor-vehicle crankshaft, which extends the whole length of the engine and must therefore possess considerable rigidity. The timing-drive for the engine valve mechanism is taken from the front end of the crankshaft, as is the pulley and belt drive for the engine auxiliaries, such as the cooling fan and the alternator for the electrical system. Attached to the rear end of the crankshaft is the engine flywheel.

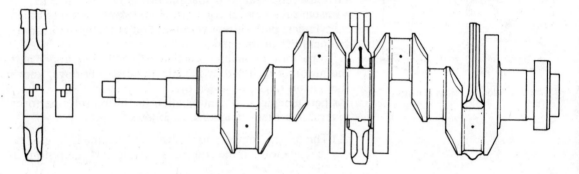

Figure 2.4 Crankshaft and bearings (Alfa Romeo)

The crankshaft is supported radially in the crankcase by a series of bearings, known as the engine 'main bearings'. Each main bearing is divided into two half-liners, similar to the big-end bearings, and again to allow assembly around the journals of the one-piece construction crankshaft.

2.1.5 Cylinder head and gasket   The functions of the cylinder head may be listed as follows:

(1) It must provide a closure or chamber for the upper part of each cylinder, so that the gas pressure created by the combustion process is constrained to act against the piston.

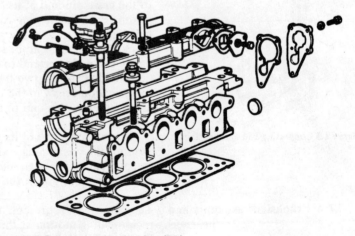

Figure 2.5 Cylinder head and gasket (BL)

(2) Associated with (1) is the need to incorporate a gas porting system with inlet and exhaust valves, as well as a platform upon which to mount their operating mechanism. Provision must also be made for a screwed boss to retain the sparking-plug.

(3) Similar to the cylinder block, the head must form a jacket that allows liquid coolant to circulate over the high-temperature metal surfaces.

(4) It is required to contribute to the overall rigidity of the engine structure and maintain a uniform clamping pressure on its sealing gasket with the cylinder block.

The sealing gasket is generally known as the 'cylinder head gasket' or simply 'head gasket'. In liquid-cooled engines the function of the cylinder head gasket is to seal the combustion chambers and coolant and oil passages at the joint faces of the cylinder block and head. The gasket is therefore specially shaped to conform to these openings, and is also provided with numerous holes through which pass either the studs or the set bolts for attaching the cylinder head to the block.

Historically, the cylinder head has not always been made detachable from the cylinder block. It was not until the early nineteen-twenties when, as mentioned previously, there was a general trend towards simplification of design which led to the abandonment of the cylinder block with integral head. This change facilitated both the production and the servicing of the motor vehicle engine, although it did introduce the risk of incorrect tightening-down of the cylinder head, which at least can cause joint trouble and, at worst, distortion of the cylinder bores.

2.1.6 Valve train and timing drive

The overall function of the valve train and timing drive is to provide first for the admission and then the retention of the

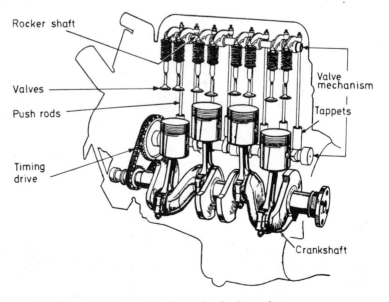

Rocker shaft

Valves

Push rods

Timing drive

Valve mechanism

Tappets

Crankshaft

Figure 2.6 Valve train for a push-rod, overhead-valve engine

combustible charge within the cylinder, and finally for the release of the burnt gases from the cylinder, all in synchronism with the motion of the pistons.

To perform this sequence of events in accordance with the requirements of the four-stroke cycle, a 'cam-and-follower' mechanism driven at one-half crankshaft speed is used to operate the engine inlet and exhaust valves. There are several different methods of operating the valves from the cam-and-follower mechanism, but in all cases it is necessary for one or more 'camshafts' to be driven from the front end of the crankshaft by what is termed the 'timing drive', and which will be examined in detail at a later stage.

It is, of course, the engine valves themselves that actually perform the functions of admitting the air and fuel charge before its compression, sealing it in the cylinder during compression and combustion, and then releasing the burnt gases after their expansion. In conventional practice, one inlet and one exhaust valve serve each cylinder, and these are mounted in the cylinder head combustion chambers. Valve springs are fitted to ensure that the motion of the valves and their operating mechanism follows faithfully that intended by the cams, and also to maintain adequate sealing pressure when the valves are closed.

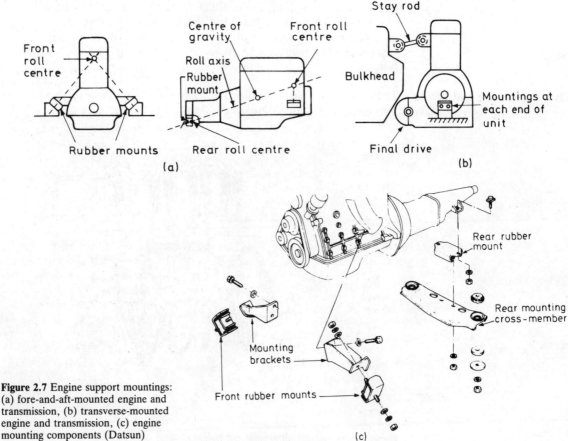

Figure 2.7 Engine support mountings: (a) fore-and-aft-mounted engine and transmission, (b) transverse-mounted engine and transmission, (c) engine mounting components (Datsun)

2.1.7 Engine support mountings

It has long since been established practice to employ a 'three-point' mounting system for the engine together with gearbox. This superseded the earlier 'four-point' mounting system not only to relieve the engine unit of all strain that could then arise from deflections of the car chassis frame, but also to absorb engine torque fluctuations more effectively by virtue of the greater mounting flexibility. The torque fluctuations induced by the power impulses from each cylinder are a major source of engine vibration, their effect being most pronounced at idling speed.

The engine support mountings almost invariably feature rubber as their spring medium, since this material is highly resilient when loaded in shear. If a rubber block is simply loaded in compression it inevitably bulges sideways, because it is deformable as opposed to being compressible, so the effect is to increase the area of rubber under load and hence reduce its flexibility as a spring. In actual practice, a compromise is often sought by loading the rubber partly in compression and partly in shear by inclining the rubber mountings. Another advantage of using rubber, rather than steel, for the spring medium is that transmission of sound through the mountings is less, since there is no metal-to-metal path for the sound to travel along.

2.2 SINGLE- *v* MULTI-CYLINDER ENGINES

Every new engine must be designed with a specific type of service in view, which then determines its general characteristics. Important among these for the car is smooth and efficient operation over a wide range of speeds and loads. Herein lies the explanation why no manufacturer lists a single-cylinder engine and very few produce an engine with less than four cylinders, the following considerations being pertinent.

2.2.1 Smoothness considerations

With a single-cylinder engine operating on the four-stroke cycle, it will be recalled that only one power impulse occurs for every two revolutions of the crankshaft. The fluctuations in crankshaft torque of a single-cylinder engine would, therefore, be quite unacceptable in motor-vehicle operation. Hence, the greater number of cylinders used, the shorter will be the interval between the power impulses and the smoother will be the flow of torque from the engine.

2.2.2 Mechanical considerations

It will be evident from the explanation of the factors governing engine power given in §1.5.8, that the power output obtainable from a single-cylinder engine of realistic dimensions and running at a reasonable speed is unlikely to be sufficient for motor-vehicle, as distinct from motor-cycle, requirements. This is because a practical limit is set on individual cylinder size by dynamic factors. Namely, the inertia forces created by accelerating and decelerating the reciprocating masses comprising the piston assembly and the upper portion of the connecting rod.

If an unusually large, and consequently heavy, piston were adopted in a single-cylinder engine intended for high-speed operation, the dynamic effects could be such as to increase the magnitude of the inertia forces to a level that at least would make engine imbalance unacceptable and, at worst, prove mechanically destructive. This is partly because the inertia forces are proportional to the cube of the piston mass (e.g. doubling the piston mass will cause the inertia forces to become eight times as great), and partly because they also vary as the square of the engine speed (e.g. doubling the engine speed will cause them to become four times as great).

2.2.3 Thermal considerations

Another problem arising from the use of an unduly large cylinder bore is that cooling of the piston and valves can be seriously impaired and this may lead to their failure from thermal overstressing. If this is to be avoided it would be necessary to incorporate special features of design, such as oil-cooling of the piston, which is often practised in marine diesel engines with very large cylinder bores.

For this and the other reasons just stated, it has therefore become established practice for the displacement of an engine to be shared among multiple 'small' cylinders, rather than a single 'large' cylinder. To this must be added the proviso that an excessively large number of cylinders increases the friction losses in an engine, quite apart from the extra complication making it more costly to build and maintain.

2.3 ARRANGEMENTS OF ENGINE CYLINDERS

Once the displacement and number of cylinders have been decided in relation to the required performance characteristics of a new engine, the next consideration is how the cylinders are to be arranged. In cars they may be arranged in three different ways, each with its own advantages and disadvantages.

2.3.1 In-line cylinders

As would be expected, all the cylinders are mounted in a straight line along the crankcase, which thus confers a degree of mechanical simplicity. Such engines are now produced with any number of cylinders from two to six. The single bank of

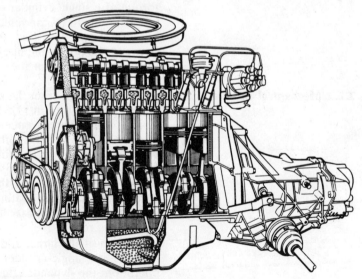

Figure 2.8 An interesting in-line five-cylinder engine (Audi)

cylinders may be contained in either a vertical, or an inclined, plane with the latter sometimes being referred to as a 'sloper' or 'slant' engine. For this particular mounting of an in-line cylinder engine, the advantages usually claimed include a reduction in overall installation height and improved accessibility for routine servicing.

With the exception of the in-line two-cylinder or 'parallel twin' engine, this arrangement of cylinders provides generally satisfactory balance in respect of the reciprocating parts, especially in six-cylinder versions. Apart from space requirements, a mechanical limitation is placed on the acceptable length of an in-line cylinder engine, because of the difficulty in controlling 'torsional vibrations' of the crankshaft. This is a topic that will be discussed at a later stage.

2.3.2 Horizontally-opposed cylinders

Horizontally-opposed engines have their cylinders mounted on the crankcase in two opposite banks and are sometimes referred to as 'flat' or 'boxer' engines. They are typically produced in two-, four- and six-cylinder versions. The main advantages usually claimed for them include inherently good balance of the reciprocating parts, a low centre of gravity that contributes to car stability, and a short engine structure. It is the latter feature that makes this arrangement of cylinders particularly suitable both for front-wheel drive and rear-engined cars, since the engine can be mounted either ahead of or behind the driven wheels with the minimum of overhang. By virtue of its low overall height, the horizontally-opposed engine can readily allow a sloping bonnet line in front-engined cars and also provide additional space for stowing luggage above it in rear-engined cars. Furthermore, it lends itself admirably to air-cooling because with an in-line cylinder arrangement it is difficult to get the rear cylinders to run as cool as the front ones, unless the engine is installed transversely.

Figure 2.9 A modern horizontally-opposed four-cylinder engine (Alfa Romeo)

The disadvantages associated with horizontally-opposed cylinders include the need for lengthy intake manifolds if a central carburettor is used, the duplication of coolant inlet and outlet connections in the case of liquid cooling, and much reduced accessibility for the cylinder heads and valve mechanism. Its greater width can also impose restrictions on the available steering movements of the wheels.

2.3.3 V-formation cylinders

With V engines the cylinders are mounted on the crankcase in two banks set at either a right angle or an acute angle to each other. They may be produced in four-, six-, eight- and occasionally twelve-cylinder versions. Where a V-cylinder arrangement has been adopted in preference to mounting the cylinders in line, it has usually been in the interests of providing a more compact and less heavy engine. In particular, the overall

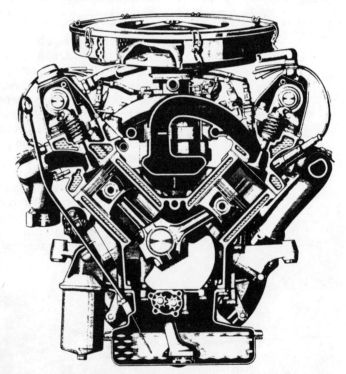

Figure 2.10 A high-performance V-eight-cylinder engine (Mercedes–Benz)

length of the engine can be appreciably reduced, so that both the structure and the crankshaft can be made more rigid. The former is thus better able to accept greater combustion loads and the latter is less prone to torsional vibrations. For typically large-displacement V-cylinder engines, the inherently wider cylinder spacings ensure adequate size of coolant passages, both around the cylinder walls and the hot exhaust valve regions in the cylinder heads.

On the debit side, the V-cylinder arrangement generally presents a more difficult balancing problem and also demands a more elaborate intake manifold from a central carburettor. In common with horizontally-opposed cylinder layouts, V-cylinder engines tend to be more costly to produce, since there are more surfaces to machine and some duplication of structural features.

2.4 CYLINDER AND CRANKTHROW ARRANGEMENTS

The arrangement of the crankthrows in relation to the disposition of the engine cylinders and their number is determined by the following and sometimes conflicting considerations:

(1) Acceptable engine balance
(2) Equal firing intervals.

2.4.1 Acceptable engine balance

Theoretically, a perfectly-balanced engine is one which, when running and 'suspended in space' from its centre of gravity, would exhibit no vibratory movements whatsoever. In reality, of

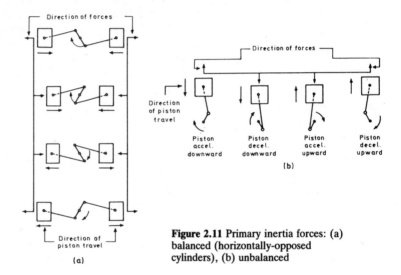

Figure 2.11 Primary inertia forces: (a) balanced (horizontally-opposed cylinders), (b) unbalanced

course, the reciprocating engine can never be perfectly balanced. Apart from any rotation imbalance, there are also the inevitable torque irregularities, although as stated previously these can be minimised by the use of more than one cylinder.

Multiple cylinders further allow a much better standard of general engine balance, provided that the choice of a particular arrangement and number of cylinders takes into account the presence of what are termed 'primary inertia forces' and 'secondary inertia forces'. The effect of these forces, which act in the reciprocating sense, will now be explained in simple qualitative terms.

Primary inertia forces: These arise from the force that must be applied to accelerate the piston over the first half of its stroke, and similarly from the force developed by the piston as it decelerates over the second half of its stroke. When the piston is around the mid-stroke position it is then moving at the same speed as the crankpin and no inertia force is being generated.

For an engine to be acceptable in practice, the arrangement and number of its cylinders must be so contrived that the

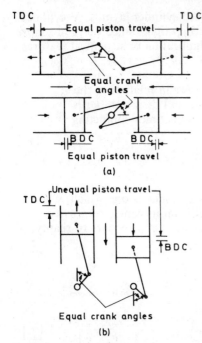

Figure 2.12 Secondary inertia forces (a) balanced (horizontally-opposed cylinders), (b) unbalanced

primary inertia forces generated in any particular cylinder are directly opposed by those of another cylinder. Where the primary inertia forces cancel one another out in this manner, as for example in an in-line four-cylinder engine with the outer and inner pairs of pistons moving in opposite directions, the engine is said to be in 'primary balance'.

Secondary inertia forces: These are due to the angular variations that occur between the connecting rod and the cylinder axis as the piston performs each stroke. As a consequence of this departure from straight-line motion of the connecting rod, the piston is caused to move more rapidly over the outer half of its stroke than it does over the inner half. That is, the piston travel at the two ends of the stroke differs for the same angular movements of the crankshaft. The resulting inequality of piston accelerations and decelerations produces corresponding differences in the inertia forces generated. Where these differing inertia forces can be both matched and opposed in direction between one cylinder and another, as for example in a horizontally-opposed four-cylinder engine with corresponding pistons in each bank moving over identical parts of their stroke, the engine is said to be in 'secondary balance'.

It is not always practicable for the cylinders to be arranged so that secondary balance can be obtained, but fortunately the vibration effects resulting from this type of imbalance are much less severe than those associated with primary imbalance and can usually be minimised by the flexible mounting system of the

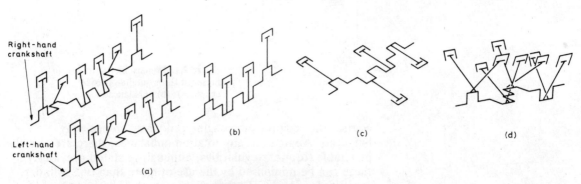

Figure 2.13 Cylinder and crankthrow arrangements: (a) in-line six-cylinder, (b) in-line four-cylinder, (c) horizontally-opposed four-cylinder, (d) V-eight-cylinder

engine. This is confirmed by the long-established and popular in-line four-cylinder engine, which possesses primary balance but lacks secondary balance.

2.4.2 Equal firing intervals

The arrangement of the crankthrows is also determined by the requirements for even firing intervals of the cylinders and for spacing the successive power impulses as far apart as possible along the crankshaft, so as to reduce torsional deflections or twisting effects. For any four-stroke cycle engine the firing intervals must, if they are to be even, be equal to 720° divided by the number of cylinders.

For in-line four-cylinder engines the first and fourth

crankthrows are therefore indexed on one side of the crankshaft and the second and third throws on the other side. The 'firing order' of these engines, numbering from the front, may then be either 1–3–4–2 or 1–2–4–3 at 180° intervals. Similarly, in the case of in-line six-cylinder engines, the crankthrows are spaced in pairs with an angle of 120° between them. Hence, the first and sixth crankthrows are paired, as are the second and fifth, and, likewise, the third and fourth. The firing order may then be such that no two adjacent cylinders fire in succession; that is, either 1–5–3–6–2–4 or 1–4–2–6–3–5 at, of course, 120° intervals.

For V and horizontally-opposed cylinder engines, the firing orders follow a sequence similar to those of in-line engines, but the crankthrows are so disposed that firing alternates between the cylinder banks as it proceeds to and fro along the crankshaft.

The numbering sequence for the cylinders of motor vehicle engines has long been defined by various standards, both international and British Standards. Stated simply, the cylinders of an in-line engine are numbered consecutively beginning from the nose end of the crankshaft. For V and horizontally-opposed engines, the cylinders are numbered consecutively, starting with the left-hand bank and then continuing with the right-hand bank, beginning from the nose end of the crankshaft in each case.

3 Cylinder block, crankcase and head

3.1 CYLINDER BLOCK AND CRANKCASE CONSTRUCTION

Cylinder blocks may be of either the 'closed-deck' or the 'open-deck' variety. The former construction represents long-established practice and resembles a deep box-like enclosure for the cylinder barrels that also serves as a coolant jacket. Transfer ducts are provided in the top face or closed deck of the cylinder block, so as to permit the circulation of coolant to the cylinder head. With the open-deck construction, the cylinder barrels are

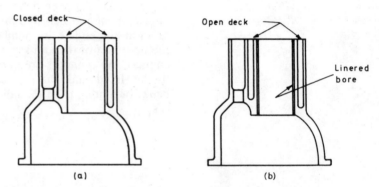

Figure 3.1 Cylinder block construction: (a) closed-deck, (b) open-deck

free-standing in that they are attached only to the lower deck of the cylinder block. By dispensing with a continuous top face, this type of construction makes for a less complex cylinder block casting and facilitates inspection of the coolant jacket interior for accumulated deposits. To perform the latter operation on a closed-deck cylinder block requires the addition of detachable cover plates.

3.1.1 Cylinder bores

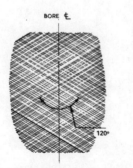

Figure 3.2 Honing pattern for cylinder bore

Clearly, the cylinder bores constitute the most important feature of the cylinder block. Since they act as a guide and a sealing surface for the sliding piston and rings, their accuracy of machining must be such as to minimise any out-of-roundness and taper effects, and to ensure that they are truly at right angles to both the crankshaft and the top deck of the block.

The cylinder bores must also be given a carefully controlled surface finish, because too rough a surface would cause wear, and a too smooth a surface would hinder the running-in process. A suitable surface finish is usually obtained by final honing to give a 'cross-hatched' finish. The question of the most suitable surface finish for new cylinder bores is one of long standing, and it is perhaps significant that an American engineer once observed that somehow the engine knows how to finish the bore better than we do!

3.1.2 Stud and screw holes

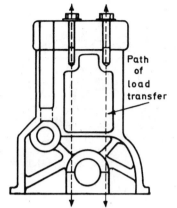

Path
of
load
transfer

Figure 3.3 Direct load transference between the studs of the cylinder head and main bearing caps

3.1.3 Crankcase construction

In both closed-deck and open-deck cylinder blocks numerous internal supporting bosses must be provided for either the studs, or the setbolts, which clamp the cylinder head to the block. Wherever practicable, these bosses are aligned with the crankcase bulkheads that support the main bearings. This is to secure a direct path of load transference between the cylinder head and main bearing caps and thus minimise bending stresses within the cylinder block and crankcase structure.

For closed-deck constructions, the internal supporting bosses for the head studs are arranged symmetrically about the cylinder walls, so that the cylinder head clamping load is as evenly distributed as possible around them. The object is, of course, to avoid any tendency towards cylinder bore distortion, which could lead to blow-by of the combustion gases and increased oil consumption. In the case of open-deck constructions, the supporting bosses for the head studs are usually disposed along the walls of the coolant jacket.

For in-line and V-cylinder engines, the most commonly used form of crankcase resembles a 'tunnel' structure, which extends downwards from the cylinder block. The roof is formed by the lower deck of the cylinder block and it is closed off at the base either by a detachable sump or a transmission housing. The crankshaft is underslung in the crankcase and supported by front, intermediate and rear main bearing bulkheads that form a series of crank chambers.

In modern practice, it is customary for the side-walls or 'skirt' of the crankcase to be extended below the axis of the crankshaft. The main bearing bulkheads may then be extended downwards, in buttress fashion, to merge with the flanged base of the crankcase skirt. Although this makes for a heavier construction, it increases the resistance to bending of the structure in the interests of engine smoothness, and also simplifies the attachment of the sump. Further, to increase rigidity, the main bearing bulkheads and the walls of the crankcase may be extensively ribbed.

For horizontally-opposed engines, the cylinder blocks are again cast integral with the crankcase, which is usually divided

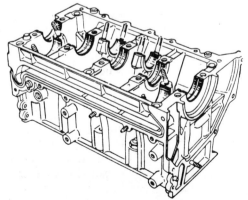

Figure 3.4 Underside view of a modern crankcase for a four-cylinder engine with five main bearings (Renault)

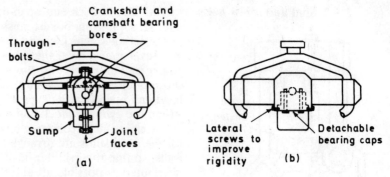

Figure 3.5 Crankcase construction for horizontally-opposed cylinders: (a) divided crankcase, (b) one-piece crankcase

on its vertical centre-line. The crankcase halves are then clamped together by through-bolts on either side of the crankshaft main bearings. A series of bolts is also fitted around the peripheral joint faces of the crankcase, it being usual for the sump to be made integral with each half.

3.1.4 Main bearing locations

For in-line and V-cylinder engines, the upper main bearing halves are carried direct in 'saddles' formed in the crankcase bulkheads, whilst detachable inverted caps of great rigidity accommodate the lower main bearing halves. These bearing caps are usually recessed into the underside of their respective bulkheads and secured to them by either studs or setbolts,

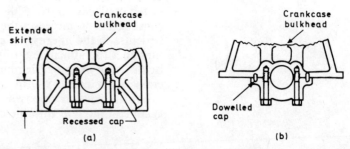

Figure 3.6 Crankcase bulkhead and main bearing cap locations

which thus support the maximum combustion loads imposed upon the crankshaft. In some designs where the crankcase-to-sump joint face is at the same level as that of the crankshaft axis, the bearing caps may be located laterally by a pair of dowels, since it is no longer expedient to recess them into the lower face of the crankcase.

Of more recent application is a form of crankcase construction that embodies a one-piece main bearing deck. This in effect integrates the main bearing caps into a single rigid structural element, rather similar to the 'bed-plate' construction found in very large diesel engines.

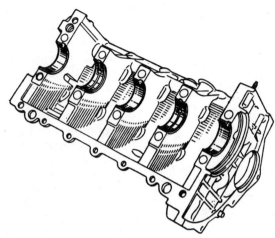

Figure 3.7 A one-piece main bearing deck (Renault)

With horizontally-opposed engines, complementary saddles for the main bearing half-liners are machined in each half of the crankcase. The main bearings cannot, therefore, be changed unless the crankcase halves are taken apart. This drawback of internal inaccessibility has been overcome in a more recent version of horizontally-opposed engine which features a one-piece construction for the cylinders and crankcase, the latter being open on its underside and provided with detachable main bearing caps and oil sump.

3.1.5 Camshaft bearing locations

A crankcase-mounted camshaft is supported either to one side of, or directly above, the crankshaft, depending on whether an in-line or a V-cylinder arrangement is used. With both locations the camshaft bearings are usually carried in webbed extensions of the corresponding main bearing bulkheads. Spanning the underside of the camshaft may be a lubrication trough, which is formed as an integral part of the crankcase. The camshaft followers or tappet barrels slide in bores machined either directly in the material of the crankcase, or in bolted-on tappet blocks. In the case of horizontally-opposed engines, the bearing bores for a central camshaft are formed similar to those for the main bearings beneath them.

3.2 CRANKSHAFT THRUST AND CAMSHAFT JOURNAL BEARINGS

Before discussing these particular bearing applications, an engineering distinction must be made between the two types of bearing. A bearing that is intended to resist a load applied perpendicular to the axis of the shaft is termed a 'radial' or 'journal' bearing. In contrast, a bearing that is intended to resist a load applied along the axis of the shaft is termed an 'axial' or 'thrust' bearing.

3.2.1 Form and materials of crankshaft thrust bearings

The crankshaft is located axially in the crankcase by 'plain' (as opposed to 'rolling') thrust bearings which restrain it against endwise movement from loading imposed mainly by the transmission system. This loading may be in a forwards direction during release of a friction clutch and in a rearwards direction when a fluid coupling is in operation.

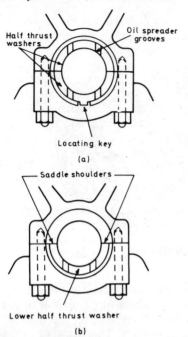

Figure 3.8 Crankshaft thrust bearings: (a) paired, (b) single

3.2.2 Form and materials of camshaft journal bearings

Crankshaft thrust bearings are generally of the same composite construction as that employed for the half-liners of the big-end and main bearings, as described later, although bronze bearings may also be used. They often take the form of separate semicircular thrust washers, which are installed either in pairs, or singly, on each side of one of the main bearing housings. Where thrust washers of smaller size are used in pairs, each lower half is keyed to the bearing cap, thus preventing both lower and upper halves from rotating once the cap is fitted. If lower-half thrust washers only are used, rotation is prevented by their upper ends abutting the joint faces of the crankcase bearing saddle. Since the thrust surfaces of the bearings must be separated by an oil film, the washers are provided with grooves or pockets to distribute the oil reaching them from the main bearing they embrace.

To accommodate both the thickness of the oil film and the thermal expansion on the parts concerned, a small clearance is provided on assembly between the thrust bearing surfaces and the adjacent contact faces of the crankshaft. This clearance must, of course, always be less than that existing between the crankweb faces and the sides of all the other main bearings, in order to relieve them of any thrust loads. The required clearance typically falls within the range 0.05–0.15 mm and is termed the 'crankshaft end-float'.

Since the loading on the camshaft bearings is generally not heavy and the operating speed of the camshaft is only one-half that of the crankshaft, the bearings are sometimes machined direct in the engine structure, especially where this is of aluminium alloy. Otherwise, established practice is to use separate steel-backed bushes that are pressed into machined bores. These bushes are lined with either a white metal alloy or

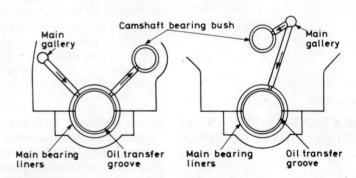

Figure 3.9 Oil supply to the camshaft journal bearings

one of the other bearing materials later referred to in connection with the big-end and main bearings. It is usual for the camshaft bearings to receive their oil supply through drillings leading from corresponding main bearings. To enable a camshaft to be inserted endwise through its bushes, the radius of its bearing journals is made slightly larger than the operating radius of its cams.

3.3 CYLINDER-BLOCK MATERIALS

The combination of cylinder block and crankcase is the single largest and most expensive component of an engine and may be produced from either cast iron or aluminium alloy.

3.3.1 Advantages and disadvantages of cast iron

Cast iron is an alloy of iron and carbon, there being two general classifications known as 'white cast iron' and 'grey cast iron'. Although both varieties have a carbon content between 2.5 and 4.0%, the difference between them is concerned with the condition in which the major portion of the carbon exists in the metal structure. Grey cast iron (so called owing to its grey rather than white appearance when fractured) is used for cylinder block and crankcase manufacture, because most of the carbon is present as flakes of graphite. This feature not only makes the material more readily machinable, but also provides a satisfactory wear- and corrosion-resistant bearing surface for the cylinder bores. The rigidity of cast iron is such that it exhibits very little tendency towards distortion under the loads and temperatures encountered in the highly-stressed engine structure. In addition, it possesses useful sound-damping properties.

Apart from its low cost, an outstanding characteristic of grey cast iron is the ease with which it can be cast into intricate shapes of thin section. Using modern casting techniques, the thickness of the cylinder-block walls can therefore be minimised to save weight, which otherwise is the only real disadvantage with the cast iron cylinder block and crankcase.

3.3.2 Advantages and disadvantages of aluminium alloy

The term 'aluminium' is generally used to describe not only the very soft and ductile commercially-pure variety, but also the numerous aluminium alloys that comprise aluminium with usually more than one element added to it. Only the latter are of interest for cylinder block and crankcase construction, since they can be made harder and more readily machinable than commercial aluminium, which tears badly and poses screw-threading problems. The main attraction of using an aluminium alloy for casting the engine structure is the saving in weight that it affords, the relative density being such that it is about one-third the weight of cast iron. On the debit side, its strength is less at about two-thirds that of cast iron. This means that metal sections have to be thickened to compensate for the lower strength, so that in reality the saving in weight is generally nearer to one-half that of the cast-iron version.

There are two main classifications for aluminium alloys: those which can be hardened by cold-working processes, and others that can be heat-treated to obtain the desired mechanical properties. It is the latter alloys of the aluminium–silicon type that find the widest application for cylinder block and crankcase manufacture, because they retain their strength at moderately high temperatures, possess good casting fluidity and are the most resistant to corrosion.

Although an appreciably lighter construction can be obtained by using aluminium alloy, its wear-resistant properties are less acceptable for the cylinder bores, so that pistons of similar material cannot run directly in them. To overcome this disadvantage either cast-iron cylinder liners, or hard-iron-coated pistons, may be employed. Other disadvantages associated with an aluminium alloy engine structure are that it is less tolerant both of careless handling, especially in respect of screw thread connections, and of accidental overheating through loss of coolant.

Historically, aluminium alloys were introduced extensively during the beginning of the aircraft industry and it is perhaps of interest to recall that one of the earliest examples of a cast aluminium crankcase was that used in the engine of the Wright brothers' first aircraft.

3.4 CYLINDER-HEAD CONSTRUCTION

In petrol engines the lower deck of the cylinder head contains the cylinder combustion chambers. Less commonly, these may be formed either within the piston heads, or by the combination of an inclined top deck for the cylinder block and specially shaped piston crowns. With the latter arrangements, the valves seat directly in the flat underside of the cylinder head.

The top deck of the cylinder head provides a platform for mounting the overhead valve mechanism and, in the case of overhead camshaft installations, it is usually extended forwards

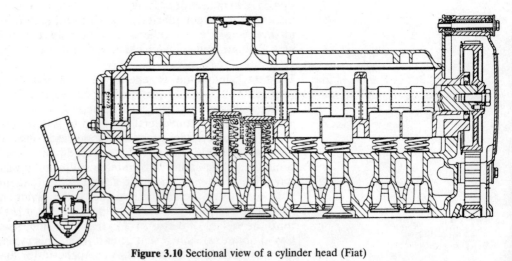

Figure 3.10 Sectional view of a cylinder head (Fiat)

to enclose the upper timing drive system. The walls of the cylinder head form the coolant jacket and provide attachment faces for the intake and exhaust manifolds. An upper continuous flange is formed by the walls for mounting the valve cover and serves to raise the sealing joint face above the level of oil draining from the valve mechanism. The coolant outlet connection at the front part of the cylinder head also serves as a housing for the thermostat, this being described in §10.5. Sandwiched between the upper and lower decks of the cylinder head and surrounded by coolant are the inlet and exhaust valve ports. These are necessarily curved and kept as short as possible, the latter to avoid excessive heat transfer both from the coolant to the induction system and from the exhaust system to the coolant. If one port serves two adjacent cylinders it is said to be 'siamesed'.

3.4.1 Valve seats
The valves may seat either directly in the material of the cylinder head or, in harder-wearing rings, inserted therein. Valve seat inserts are usually confined to engines where the cylinder head material is aluminium alloy. Modern valve seat

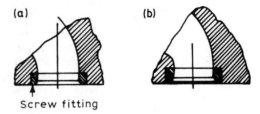

Figure 3.11 Valve seat inserts: (a) early practice, (b) later practice

inserts usually take the form of plain rings of greater depth than width, their proportions being such as to confer adequate resistance against distortion. In earlier applications, the screwed-in type of insert was sometimes used. For both types it is necessary that they be made an interference fit in the cylinder head.

Valve seat inserts for aluminium alloy cylinder heads are generally produced from a nickel alloy iron casting, which not only imparts the desired hot strength, hardness and corrosion resistance, but also has a matching coefficient of expansion.

3.4.2 Valve guides
Coaxial with the valve seatings are the valve guides, which are carried in bosses extending from inside the valve ports to the top deck of the cylinder head. Their length must be such as to present an adequate bearing surface to resist any side loading on the valve stems, and also to provide a ready path of heat transfer from the exhaust valve head. They are further arranged to project above the level of oil draining from the valve mechanism onto the top deck of the cylinder head.

Cast iron is generally the preferred material for the valve guides, which may be either made removable from the cylinder

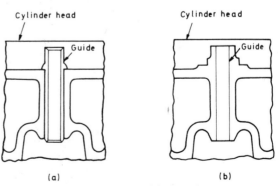

Figure 3.12 Valve guides: (a) removable, (b) integral

head or, as in American practice, cast integrally with it. Separate cast-iron guides are, of course, required with aluminium alloy cylinder heads. Removable guides are always made an interference fit in the cylinder head material, thereby assisting heat transfer from the valve to the cooling medium, which surrounds their supporting bosses.

3.5 CYLINDER-HEAD MATERIALS   The materials used for cylinder-head construction are either cast iron or aluminium alloy, as in the case of the cylinder block and crankcase. Apart from effecting a saving in weight, the greater heat conductivity of aluminium alloy is beneficial in maintaining a more uniform temperature throughout the cylinder head. This may sometimes permit the compression ratio of an engine to be raised slightly without incurring detonation or knocking. An aluminium alloy head has often been used in combination with a cast-iron block and crankcase, and it is usually considered essential for air-cooled engines to ensure efficient heat transfer.

A service difficulty sometimes encountered with aluminium alloy cylinder heads is that they can prove obstinate to remove if corrosion deposits build up around their fixing studs, and it is for this reason that some designers specify setbolts.

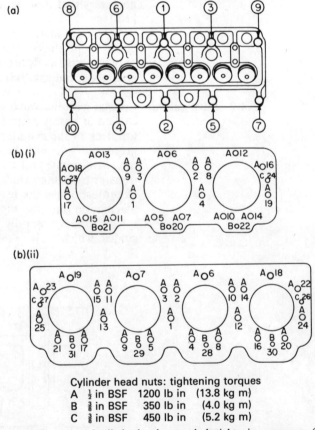

Cylinder head nuts: tightening torques
A $\frac{1}{2}$ in BSF   1200 lb in   (13.8 kg m)
B $\frac{3}{8}$ in BSF   350 lb in   (4.0 kg m)
C $\frac{3}{8}$ in BSF   450 lb in   (5.2 kg m)

Figure 3.13 Examples of cylinder head nut or bolt tightening sequence: (a) car petrol engine (Toyota), (b) commercial-vehicle diesel engines (i) 6LX and 6LXB, and (ii) 8LXB cylinder heads (Gardner)

3.6 TIGHTENING-DOWN THE CYLINDER HEAD

Before carrying out this operation it is always advisable to consult the particular manufacturer's service instructions, especially in respect of the following:

(1) Check whether the screw threads and washer faces require lubrication, and if so, the type of lubricant to be used. The point here is that the presence or otherwise of a lubricant on screw-threaded assemblies will affect the clamping load they exert for any given value of torque tightness.
(2) Establish the correct sequence of nut or setbolt tightening, the number of stages in which it is to be achieved and the final torque value to be attained. The numerical sequence of tightening specified usually involves starting from the centre and working alternately towards each end.

Any re-tightening of the cylinder head after the engine has been run should be done strictly in accordance with the manufacturer's recommendations.

3.7 INCORRECT TIGHTENING OF THE CYLINDER HEAD

The consequences of incorrect tightening of the cylinder head can prove very expensive to rectify. They may include at the least coolant and compression leaks via the head gasket and, at worst, the failure of screw-thread fastenings and distortion of the cylinder bores.

3.8 CYLINDER-HEAD GASKET

Although the mating faces of the cylinder head and block are machined smooth, flat and parallel, in reality there are always minute surface irregularities and structural deflections to be accommodated. A static seal or 'head gasket' is therefore required, which possesses the necessary degree of both plasticity

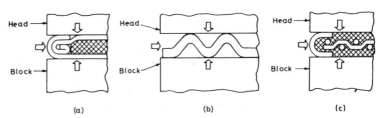

Figure 3.14 Types of cylinder-head gasket: (a) metal and asbestos layers, (b) embossed steel sheet, (c) metal-cored composition materials

(pliable) and elasticity (springy). The effect of tightening-down the cylinder head on to the block places the gasket under compression, so that sufficient friction is created between the sealing surfaces to resist extrusion of the gasket by cylinder gas pressure.

3.9 CYLINDER-HEAD GASKET MATERIALS

These must, of course, be resistant to the effects of heat and are generally produced from either copper and asbestos layers, embossed steel sheet or, as in more recent practice, metal-cored composition materials such as asbestos and plastics. Where gaskets of the first and last mentioned types surround the combustion chamber openings, their edges are sometimes protected by steel beading, which may also be required to increase sealing pressure in these regions.

3.10 CRANKCASE SUMP

This unit acts as a reservoir to store the oil that is required by the engine lubrication system. It further serves as a vessel in which any sludge, water and metal particles in the oil can settle out, and also provides an opportunity for any entrained air to escape from the oil.

The sump is either of pressed-steel construction, or produced from an aluminium alloy. In its latter form it better contributes to the rigidity of the crankcase and also assists with heat dissipation. Attachment of the sump to the crankcase is usually by means of setscrews through mating flanges, between which is sandwiched a flexible packing or gasket, as later described in §7.6.

Baffle plates are normally fitted in the sump to minimise both oil surging and agitation, the former arising from the changing motion of the car and the latter from the oil flung from the crankshaft bearings. A screwed plug is incorporated at the lowest point in the sump for draining the oil in service.

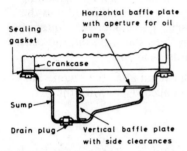

Figure 3.15 Crankcase sump

3.11 CYLINDER LINERS

During an earlier era of rapid bore wear, cylinder liners were quite widely used either as original equipment or as an overhaul feature. This was because the particular grade of iron from which they were cast centrifugally could be selected for its wear-resistant properties, rather than for the free-flowing characteristics required of an iron for casting the cylinder block and crankcase. However, later developments in the fields of piston-ring coatings, lubricating oil formulation, oil and air filtration equipment and cooling system control have all combined to minimise cylinder bore wear, so that the need for detachable liners on this score seldom arises. Cylinder liners are now generally specified either to provide a suitable wear-resistant surface for the cylinders of aluminium alloy engines, or to simplify the production of cast-iron engines by permitting an open-deck form of cylinder block.

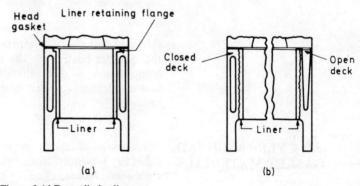

Figure 3.16 Dry cylinder liners

3.11.1 Dry cylinder liners

A detachable 'dry' liner takes the form of a plain or a flanged sleeve, the entire outer wall of which is maintained in intimate metal-to-metal contact with the cylinder block. This is of closed-deck construction and may be of either cast iron or, less commonly, aluminium alloy.

Non-detachable dry liners have been cast integrally with aluminium alloy cylinder blocks of both closed- and open-deck constructions.

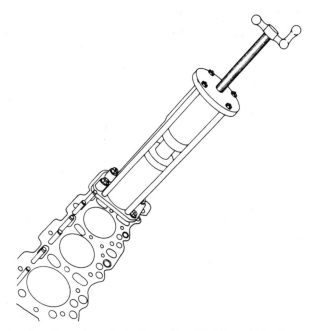

Figure 3.17 Installing a dry cylinder liner with a Flexi-Force hydraulic press (Brown Bros.)

Dry liners generally contribute to the rigidity of the cylinder block, but tend to introduce a barrier to heat flow at the adjoining surfaces. This effect is minimised where the cylinder block is made from aluminium alloy, as a consequence of its good heat conductivity.

3.11.2 Wet cylinder liners

The 'wet' type of liner always takes the form of a flanged sleeve, the outer wall of which is largely exposed to the coolant

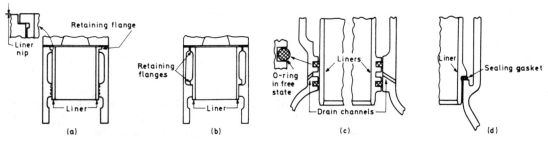

Figure 3.18 Wet cylinder liners in (a) closed-deck, and (b) open-deck cylinder blocks. Sealing arrangements for suspended wet liners, (c), and wet liner held in compression (d)

in the cylinder jacket. It may be incorporated in both closed- and open-deck cylinder block constructions. Clearly, the wet cylinder liner is better-cooled than the dry type and can more easily be renewed when worn, as will be explained later. It contributes little to the rigidity of the cylinder block, however, and there is always the possibility that coolant leaks into the crankcase may occur.

3.12 WET CYLINDER LINER INSTALLATIONS

Two distinct methods of locating wet liners may be used, according to whether they are being installed in closed- or open-deck cylinder blocks, as follows:

(1) *Closed-deck:* The cylinder liner is provided with a top flange only and is suspended through the coolant jacket from where it is clamped between the cylinder head and the upper deck of the cylinder block.

(2) *Open-deck:* Here the cylinder liner must be provided with a top and a lower flange and it is held in compression within the coolant jacket between the cylinder head and the lower deck of the cylinder block.

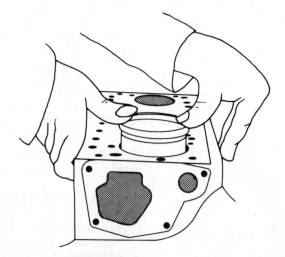

Figure 3.19 Inserting a wet cylinder liner by hand (Perkins)

An advantage of the first arrangement is that the cylinder block is relieved of stresses that would otherwise be imposed by the axial expansion of the liner upon heating. With the second arrangement, the less intrusive top flange generally permits of better cooling around the upper part of the liner.

3.12.1 Sealing arrangements

For closed-deck cylinder blocks, a pair of oil-resistant (synthetic) rubber O-ring seals encircles the lower part of the liner and are deformed into grooves where it passes freely through the lower deck of the cylinder block. The sealing rings may be grooved into either the cylinder block or the liner itself, and a third unfilled groove between them communicates with a drilling in the block that leads to atmosphere. This drilling serves as a drain channel for any coolant and, similarly, oil that may have seeped past the top and bottom sealing rings, respectively.

With open-deck cylinder blocks, a compression sealing gasket is generally used between the flange towards the bottom of the liner and its seating in the lower deck of the block.

In both types of liner installation the cylinder head gasket completes the sealing arrangements for the top end of the liner.

3.13 CYLINDER LINER SERVICING CONSIDERATIONS

It will be evident from the previous descriptions that the dry liner is usually (although not invariably) made an interference fit in the cylinder block. Typically, the block is bored out to provide an interference fit of 0.06–0.09 mm between the cylinder and the liner, which will then need to be lubricated and pressed in under a load of about 2000–3000 kg. To avoid any possibility of liner bore distortion, the cylinder block studs are usually refitted before the liner bore is honed to final size.

In contrast, wet liners are generally made what is termed a 'slip-fit' in the cylinder block. A typical cylinder liner to block clearance would be 0.05–0.15 mm. Even so, a manufacturer may recommend that the cylinder block be pre-heated, so that there is no hindrance to correct insertion of the liners and seals. There is always the danger that if an engine with wet liners is cranked over with the cylinder head removed, the liners could be dragged clear of their locations by the rising pistons. The temporary fitting of retaining clamps on the liners is therefore the safest practice in these circumstances.

A final consideration in fitting flanged liners, either wet or dry, is the provision of a small amount of what is termed 'nip'. This refers to the amount the liner top flange protrudes above the top deck of the cylinder block, so as to promote an efficient gasket seal when the cylinder head is clamped onto it. In practice, a liner nip of 0.05–0.12 mm is fairly typical.

4 Piston assemblies and connecting rods

4.1 PISTON MATERIALS AND CONSTRUCTION

In low-speed engines of early design, the material from which the pistons were made was cast-iron to match that of the cylinders. With increasing engine speeds and output, however, it has long since become established practice for the pistons to be produced from aluminium alloy, materials of this type combining lightness in weight with high thermal conductivity. They have a moderate silicon content so that their mechanical strength is better maintained at high operating temperatures, which may now exceed 300°C, whilst their coefficient of thermal expansion is lower than that usually associated with aluminium alloys.

4.1.1 Structural features and nomenclature

The piston 'crown' may be either flat-topped or specially shaped in order to conform to the particular design of combustion chamber of which it forms one wall. Combustion loads are transmitted directly from the crown to the 'gudgeon-pin bosses' through intermediate supporting webs, which also facilitate the flow of heat to the encircling piston rings and thence to the cylinder walls. The 'ring belt' immediately below the crown is thus largely relieved of loads that would otherwise tend to deform its grooves. Any closing in of the grooves would, of course, prevent the free radial movement of the rings and thus impair their sealing ability.

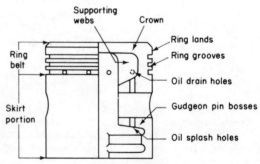

Figure 4.1 Piston construction and nomenclature

The main part of the piston below the ring belt is termed the 'skirt', and this is made as close-fitting as practicable in the cylinder, thereby ensuring quiet operation and the maintenance of the rings at their most favourable attitude to the cylinder wall. Futhermore, the skirt must present an adequate bedding area to the cylinder wall, not only to minimise contact pressure, but also to assist with heat dissipation. It should be appreciated though that the piston skirt is not normally in direct contact with the cylinder wall, but is separated from it by a film of oil.

4.2 TYPES OF PISTON SKIRT

Modern practice favours the use of a 'solid' skirt piston of rigid construction, because of the high combustion loads now encountered. Basically, its advantages are that it can be made thinner in section to withstand a given loading, so that it affords a saving in weight. It does, however, need a good deal of modification to provide acceptable expansion control of the skirt, as will be explained later. Another type of skirt which was once widely used is that known as the 'split' skirt. This incorporated a near-vertical slot extending from the centre of an upper horizontal slot down to the base of the skirt on the 'non-thrust' side of the piston. Here it should be explained that the 'thrust' side of the piston reacts the side force arising from the angular motion of the connecting rod on the power stroke, while its 'non-thrust' side reacts the lesser side forces on the compression and exhaust strokes. The split skirt piston was originally introduced to provide quiet running and, by virtue of its skirt flexibility, to accommodate a certain degree of cylinder bore distortion where this was prone to occur.

4.3 EXPANSION CONTROL FOR PISTONS AND RINGS

Although the aluminium alloys chosen for piston material expand less when heated than most others, they nevertheless expand at nearly twice the rate of the cast iron used for most engine cylinders. In the absence of special expansion control features, this relative difference in expansion rates would result in seizure of the piston when hot, and could be avoided only by tolerating a too slack fitting in the cylinder when cold that would result in 'piston slap'.

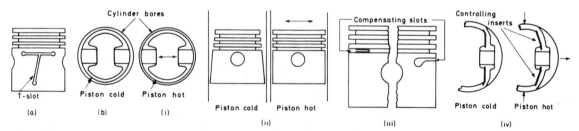

Figure 4.2 Methods of expansion control for pistons. (a) Split-skirt piston with compensating T-slot. (b) Solid-skirt pistons with (i) ovality, (ii) taper, (iii) compensating slots, and (iv) controlling inserts

A basic method of controlling thermal expansion is to machine the piston to a special form, which is both oval in contour and tapered in profile. The ovality is such that when the piston is cold, the minor axis of the skirt lies in the direction of the gudgeon pin. When the piston is hot the skirt then assumes a circular shape, because of the greater expansion occurring in the mass of metal comprising the gudgeon pin bosses.

The direction of taper allows for additional clearance when cold near the top of the skirt, since this part ultimately attains a greater running temperature and thus expands more than the cooler running lower portion. Immediately below the ring belt, where the skirt temperature is greatest, the degree of taper may be intensified to produce an overall 'barrelled ' shape. Similarly, the clearances around the ring 'lands' are progressively increased towards the piston crown, thereby avoiding contact when hot with the cylinder wall.

In order that the piston can be made as close fitting as possible in the cylinder, either compensating slots or controlling inserts, or both, may be incorporated in its construction. With split skirt pistons, circumferential expansion of the skirt is simply absorbed by temporary closing in of the near-vertical compensating slot. For solid skirt pistons it is usual for part-circumferential slots to be located within or, in some designs,

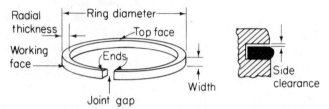

Figure 4.3 Piston ring nomenclature

beneath the oil control ring groove. The angular disposition of these horizontal slots is such that the flow of heat from the piston head is diverted from the thrust faces of the skirt, thereby reducing expansion across them. In American and German practice, alloy steel inserts with a lower coefficient of expansion than the piston material bow outwards when heated and thus restrain expansion of the skirt across the thrust axis.

4.3.1 Methods used for piston rings

A piston ring must be provided with a gap at one point on its circumference for three reasons, as follows:

(1) So that it can be expanded over the piston head and then released into its groove.
(2) To allow it to be compressed into the cylinder and thus exert an initial sealing pressure, which is then greatly augmented by gas pressure when the engine is running.
(3) To accommodate circumferential expansion of the ring when it is hot.

With regard to (3), although the coefficients of expansion of the cast iron and alloy steels from which the piston rings are made relate to that of the cylinder material, the circumferential expansion of the rings is greater because they attain a higher mean temperature, otherwise there would be no heat flow from one to the other. It is also for this reason that a piston ring which maintains a uniformly distributed pressure against the cylinder when cold, is not likely to do so when hot, and this has therefore to be taken into account during manufacture.

4.4 FITTING CLEARANCES FOR PISTONS AND RINGS

Since the normal operation of the pistons and their rings accounts for about 75% of the total friction losses in an engine, the fitting clearances relating to the combination of piston, rings and cylinder should always be in accordance with the manufacturer's specification. For this reason the following data are intended only as a general guide:

(1) *Piston clearance:* This must always be sufficient to maintain an oil film and thereby prevent scuffing or seizure of the piston in the cylinder bore. It rarely exceeds 0.05 mm measured at the top of the skirt across the thrust axis. If an engine is fitted with wet cylinder liners, it is usual for matched liners and pistons to be made available in service.

(2) *Piston ring closed gap:* To compensate for circumferential expansion this should usually be not less than three-thousandths of the cylinder bore diameter, and is measured with the ring installed in its cylinder. In practice, a gap of between 0.30 and 0.35 mm is likely to be specified.

(3) *Piston ring side clearance:* To allow for free radial movement of the ring in its groove this should usually be not less than 0.035 mm for cylinder diameters of between 50 and 100 mm.

4.5 TYPES OF PISTON RING

There are two basic types of piston ring used in petrol and diesel engines, these being designated 'compression' and 'oil control', and in each case their nomenclature is most conveniently presented by illustration.

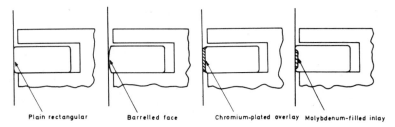

Plain rectangular   Barrelled face   Chromium-plated overlay   Molybdenum-filled inlay

Figure 4.4 Top compression rings

4.5.1 Petrol engine piston rings

For modern petrol engines two compression rings and one oil control ring are used on each piston. In earlier practice it was customary for more than three rings to be used. For example, the fairly low compression ratio engines used in Rolls–Royce cars of the late nineteen-twenties originally had five-ring pistons. The basic reason why fewer rings can now be used, therefore lies in the higher mean cylinder pressures that better augment the sealing action of the rings. Various forms of compression ring may be specified, the differences between them lying in their cross-sectional shapes and the application of wear-resistant surface treatments. Top compression rings are usually of plain rectangular section, their inner and outer edges being slightly chamfered to prevent sticking in the groove. The working surface of the ring may also assume a 'barrel' form, instead of being flat and parallel to the cylinder wall, so that it is better able to accommodate any slight piston rock where the skirt length may be limited.

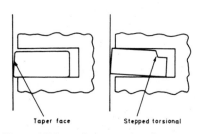

Taper face   Stepped torsional

Figure 4.5 Second compression rings

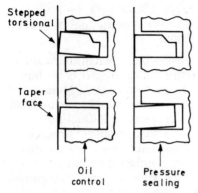

Figure 4.6 Action of second compression rings

The second compression ring principally serves to reduce the pressure drop across the top ring, and it can with advantage be made relatively more flexible so as also to assist oil control. Various departures are therefore made from the basic rectangular cross-section, the main object being to compensate for torsional deflection of the ring under combustion pressure, so that top edge contact with the cylinder wall is avoided. Otherwise, the ring tends to pump oil towards the combustion chamber and therefore opposes the action of the oil control ring. To reverse this effect, 'taper-faced' and 'stepped-torsional' rings are widely used, the two features being combined in some designs.

The basic 'slotted' type of oil control ring is simply a modification of the plain rectangular-section compression ring. It takes the form of an outward-facing channel section from which collected oil escapes through slots machined radially to the back of the ring. The oil then returns to the sump via communicating holes drilled through the piston wall at the back of the groove. A fairly high radial pressure is exerted against the cylinder wall by virtue of the narrow working surfaces of the ring, so that any tendency for it to ride over the oil film is counteracted.

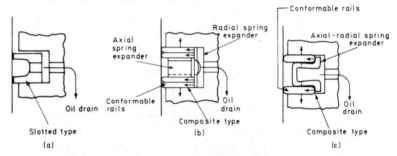

Figure 4.7 Oil control rings: (a) slotted type, (b) and (c) composite types

Oil control requirements have become more exacting for modern engines because the use of high compression ratios increases the depression acting on the rings during the induction stroke, especially on the overrun with a closed throttle. To improve oil control under these conditions various forms of composite rings are commonly used. They generally feature two or more independent working faces, comprising flexible rails with rounded wiping edges, which act in conjunction with a spring expander. This presses the rails axially against the sides of the ring groove and radially against the cylinder wall. Since the flexible rails can act independently of one another, they do not lose their effectiveness in the presence of any piston rock.

4.5.2 Diesel engine piston rings   The piston and rings of the diesel engine have to perform an altogether more arduous duty than those in the petrol engine, and must also provide much longer life in order to satisfy the requirements of commercial vehicle operators. Apart from differences in piston design that will be described at a later stage, the piston ring combination generally includes an

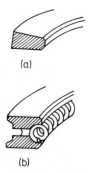

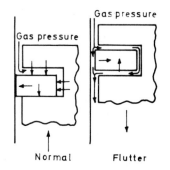

Normal   Flutter

Figure 4.8 Diesel engine piston rings, (a) wedge compression ring, (b) conformable control ring

Figure 4.9 Piston ring sealing action

additional compression ring, making three in all, and one oil-control ring. The compression rings follow closely upon current petrol engine practice, an exception being the taper-sided or 'wedge' ring that was intended to prevent ring sticking. This was once a prevalent complaint with diesel engines and is now much reduced by modern detergent oils, although it can still be caused by faulty operation of the fuel injectors. The idea behind the wedge ring was that any slight rocking of the piston would alter the ring side clearance and thus discourage the build-up of carbon deposit.

Diesel engine oil-control rings are generally more rugged than the composite type used in petrol engines and take the form of a spring-loaded conformable ring. This is basically a thinner slotted oil control ring with a backing spring.

4.6 PISTON-RING FAULTS AND THEIR CAUSES

Piston ring misbehaviour in an engine is indicated by loss of cylinder compression, poor oil control and noisy operation. These complaints can result from wear, sticking or breakage of a ring, although the fault may not necessarily lie with the ring itself, and can be summarised as follows:

(1) *Abrasive wear:* This has been much reduced by modern air filtration equipment and closed-circuit crankcase ventilation, which prevents air-borne dust particles from entering the cylinders and the crankcase, thereby protecting the compression and oil-control rings, respectively. Without some abrasive wear the piston rings would never, of course, bed-in at all!

(2) *Scuffing wear:* A definite explanation for this particularly severe type of wear has long been the subject of engineering research (since it can also occur, for example, between gear teeth), but it is generally thought to result from the formation and tearing of tiny welds on the sliding surface of the ring, under the effects of pressure and temperature. If it occurs it does so rather suddenly, typically during the running-in of an engine.

(3) *Corrosive wear:* This is attributed to chemical attack by the products of combustion and occurs when these are able to condense on the ring surfaces at low temperatures. It is now

minimised by better regulation of the cooling system, so that the engine warms up more rapidly and takes longer to cool down, and also by applying various surface treatments to the ring face.

(4) *Sticking:* Reference has already been made to this condition and its prevention by modern detergent oils in connection with diesel engines and suffice it to say here that it was similarly once prevalent in petrol engines. It occurs when the ring groove temperature becomes excessive and causes breakdown of the lubricating oil reaching the ring, so that solid products build up in the groove and prevent free radial movement of the ring.

(5) *Breakage:* Apart from the breakage that can result from the sticking of a ring, another condition known as 'ring flutter' may cause the breakage of rings in all cylinders if an engine is persistently over-revved. Following experiments made some 30 years ago by Dr P de K Dykes, it is now recognised that if the reciprocating inertia force acting upwards on the ring exceeds the gas pressure that is forcing it downwards and outwards, then during the end of the compression stroke and the beginning of the power stroke, the ring will lose contact with the lower side of its groove. The consequent release of gas pressure from behind the unseated ring results in it collapsing radially inwards and this can lead to failure.

4.7 PISTON-RING MATERIALS

Compression and slotted-type oil control rings are produced from cast iron, because this material provides a satisfactory wear-resistant surface and retains its elasticity at high temperatures. Composite oil control rings for petrol engines utilise thin steel rails with chromium-plated working edges, whilst in the spring-loaded conformable ring used in diesel engines, a cast-iron slotted ring is expanded by a stainless steel backing spring.

Since the top ring is subject not only to the immediate effects of combustion pressure and temperature, but must also withstand abrasive and corrosive conditions, a wear-resistant coating is now generally applied to its face. This may comprise either a hard chromium-plated overlay or a molybdenum-filled inlay. The former treatment was actually first applied to piston rings of military vehicle engines during the African campaign in World War II, and is especially beneficial in resisting abrasive and corrosive wear. Molybdenum inlay is a more recent development, and is particularly resistant to scuffing wear. Similar benefits are, of course, conferred on the cylinder wall itself.

4.8 METHODS OF GUDGEON-PIN LOCATION

These vary in accordance to whether a 'semi-floating' or a 'fully floating' gudgeon pin is used, as follows:

(1) *Semi-floating gudgeon pin:* This is held rigidly in the connecting rod eye and oscillates only in the piston bosses. Current practice is for the gudgeon pin to be retained by an interference fit, rather than by clamping it in a split eye with

a pinch-bolt as in earlier designs. The latter method introduced a discontinuity into the connecting rod eye that could be a source of weakness.

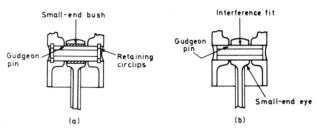

Figure 4.10 Methods of gudgeon-pin location: (a) fully floating, (b) semi-floating

(2) *Fully floating gudgeon pin:* In order to prevent it from escaping sideways and contacting the cylinder wall, the fully floating gudgeon pin must be located axially in the piston bosses. This is because it is free to oscillate both in the connecting rod eye and the piston bosses at normal operating temperature. Location is usually provided by spring retaining rings or 'circlips', which are expanded into grooves near the outer end of each boss and thus act as removable shoulders.

4.9 SMALL-END AND BIG-END BEARINGS

Although the gudgeon pin and small-end bearing directly react the combustion load, they are nevertheless made appreciably smaller in diameter than the crank pin and big-end bearing. This difference in size is not simply to accommodate the small-end of the connecting rod within the piston, but can be justified by the small-end bearing benefiting from a much reduced share of the total reciprocating and rotating inertia forces created by the connecting rod and piston.

4.9.1 Form and materials of the small-end bearing

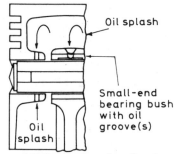

Figure 4.11 Lubrication of small-end bearing

Where a fully floating type of gudgeon-pin is used, a separate bearing bush is pressed into the eye of the connecting rod. Since this bearing is difficult to lubricate effectively, because of its oscillating rather than rotating motion, it must possess a high degree of durability. It is now generally of composite construction and comprises a steel backing lined with a hard lead–bronze alloy. To assist lubrication, the upper bearing surface of the bush may be grooved to form a reservoir for collecting oil entering through a drilling in the rod eye. The assembly fit of the gudgeon pin in the small-end bush tends to be critical, since too much clearance can produce a small-end tapping noise. Too little clearance when cold can result in the piston being rocked by the angular motion of the connecting rod, thereby causing a temporary piston knocking noise.

4.9.2 Form and nomenclature of big-end bearings

Since the early nineteen-thirties it has become established practice for the big-end (and main) bearings to take the form of thin, flexible half-liners, their nomenclature being most

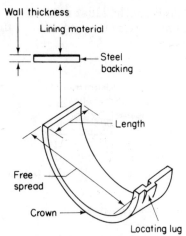

Wall thickness
Lining material
Steel backing
Length
Free spread
Crown
Locating lug

Figure 4.12 Big-end half-liner

conveniently presented by illustration. They are of composite construction and consist of a preformed thin steel backing or 'shell' to which is bonded one or more very thin layers of relatively soft bearing material. Bearings of this type therefore became known as 'thin-wall' and were originally introduced by the Cleveland Graphite Bronze Company in America. Before the advent of thin-wall bearings, the housing was often lined direct with the bearing material, which then required boring and hand scraping to achieve a satisfactory fitting. Thin-wall bearings possess several important technical advantages, including a much improved fatigue resistance, a more compact installation and better suitability to mass production requirements. Fatigue resistance concerns the ability of the bearing to withstand fluctuating loads at fairly high temperatures, and is the single most important property required of a bearing. It is achieved in the thin-wall bearing because there is less deformation taking place in the much thinner layer of bearing material.

4.9.3 Big-end bearing materials

A heavier-duty bearing material may be selected for the connecting rod big-end bearings than is used for the crankshaft main bearings, because it is subjected to centrifugal loading by the partly rotating motion of the lower part of the connecting rod. This effect is absent in the case of the main bearings supporting a counterbalanced crankshaft. The big-end (and main) bearings must not only possess adequate fatigue resistance, but also provide satisfactory wear qualities. These requirements tend to conflict in practice, since the relatively soft materials that have the best anti-wear properties are usually the least resistant to fatigue. The various bearing lining materials used can be listed as follows:

(1) *White metal alloys:* These are either tin-based or lead-based and are sometimes referred to as 'babbit' metals, although strictly speaking this description should be confined to the tin-based alloy first developed in 1839 by Isaac Babbit of Massachusetts. They contribute to a low rate of wear on the crankpins because of their good 'embeddability', which means they can readily bury any unfiltered wear particles entering the bearing oil film clearance and thus prevent scoring. The good 'conformability' of these alloys also means they can tolerate any slight misalignment and deflections that may affect a bearing. Their load-carrying capacity is, however, limited by a modest fatigue resistance that decreases rapidly with increasing temperatures.

(2) *Copper–lead mixtures:* Harder materials of this type were originally developed to meet the combination of increased loads and higher operating temperatures. They possess a higher fatigue resistance than white metal alloys, but show inferior embeddability and conformability. These disadvantages were largely overcome by plating the lining material with a very thin overlay of a softer lead-based alloy, which also improves its corrosion resistance. It is usual for copper–lead bearings to require surface-hardened crankshaft journals.

(3) *Aluminium–tin alloys:* These have come into widespread use

during the last 20 years and compare favourably with copper–lead bearings in respect of fatigue resistance. Their wear performance is usually such that a plated overlay is seldom required. Furthermore, the corrosion resistance and thermal conductivity are both high for this type of bearing. More recently, silicon–aluminium alloy has been developed as a high-duty bearing material for diesel engine application.

4.10 METHODS OF LOCATING BIG-END BEARINGS

Since these bearings are made detachable from the big-end of the connecting rod, they must be located in both the rotational and axial senses, as follows:

(1) To prevent rotational movement, the two half-liners are retained in their housing by an interference fit which, by virtue of maintaining an intimate metal-to-metal contact between liners and housing, also facilitates heat flow from the bearing. The interference fit is obtained by extending the half-liners a few hundredths of a millimetre beyond their true parting line, so that they are compressed into their housing when the bearing cap is tightened down. This difference in circumferential length between the pair of abutting liners and the closed bore of the big-end housing is known as the bearing 'nip' or 'crush'.

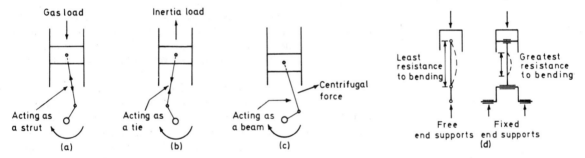

Figure 4.13 Forces acting on the connecting rod: (a) compression, (b) tension, (c) bending, (d) concentration of bending load

(2) To prevent axial movement the two half-liners incorporate lugs that register with offset notches in both the connecting rod and its cap. The latter is fitted so that its notch faces the same side as the notch in the rod.

4.11 CONNECTING-ROD FORM AND MATERIALS

Its nature of loading is such that the connecting rod is subjected to a combination of axial and bending stresses, the former arising from reciprocating inertia forces and cylinder gas pressure, and the latter from centrifugal effects. The shank of the connecting rod is provided with an I cross-section for maximum rigidity with minimum weight. Since the end-supports for the rod are 'free' in the plane of rotation and 'fixed' in the plane containing the crankpin and gudgeon-pin axes, the largest dimension of the I section is disposed in the plane of rotation to resist the greater bending effect therein. The depth of web in the I-section varies in accordance with any taper of the shank, but the flange width and web thickness remain constant over this length. To reduce stress concentrations the big-end arch and the small-end eye are merged very gradually into the shank portion of the connecting rod.

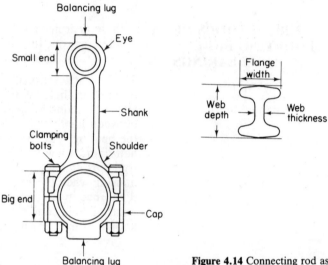

Figure 4.14 Connecting rod assembly

In some designs the big-end bearing parting line is arranged diagonally, because otherwise the width of the housing would be such that the connecting rod could not be passed through the cylinder for assembly purposes. To resist the greater tendency for the cap to be displaced sideways relative to the rod, either a serrated or a stepped joint is generally preferred for their mating faces. Hence, the securing setscrews in their clearance holes are relieved of all shear loads. Where the parting line between the rod and cap is arranged at right angles to the axis of the shank, the cap may be secured by either bolts and nuts, studs and nuts, or setscrews. They are produced from high-tensile alloy steel with special care being taken in their detail design to avoid stress raising sharp corners, which would lower their fatigue resistance. Their clamping load must always be such as to exceed the inertia forces acting on the cap.

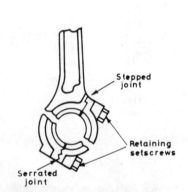

Figure 4.15 Types of angled big end

4.11.1 Connecting-rod materials

A prime requirement for a connecting rod material is that it should possess a high strength-to-weight ratio. Connecting rods are therefore either forged from a high-strength alloy steel or, as in more recent American practice, cast from a high-duty iron. With the latter method of manufacture a closer weight tolerance can be maintained than is possible with forgings.

4.12 CHECKING BEARING-LINER NIP

The following precautions should generally be observed when carrying out this operation, although it is always advisable to consult the particular manufacturer's service instructions:

(1) Ensure that the bearing seatings are absolutely clean.

(2) Check that the correct replacement bearing half-liners are being fitted.

(3) Note that the half-liners have a certain amount of 'free spread', so that they can be sprung and retained in position during assembly.

(4) Position the half-liners so that their lugs register correctly with the locating notches in the rod and cap seatings.

(5) Oil the bearing working surfaces and fit the cap the correct way round.

(6) Check that the specified bearing nip or crush is present by first tightening the cap nuts or setscrews to the torque value specified by the manufacturer, then slackening one side to finger-tight and inserting an appropriate thickness feeler gauge between the joint faces of the rod and cap. This procedure may be repeated on the other side. A bearing liner nip in the region of 0.08–0.10 mm is fairly typical. Finally, of course, the cap is retightened.

Where a check on the actual bearing clearance is called for, this should usually be about one-thousandth of the crankpin diameter. An insufficient clearance space for the oil film could lead to an excessive rise in bearing operating temperature, which lowers the fatigue resistance of its material.

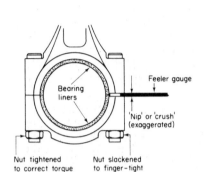

Figure 4.16 Checking bearing liner 'nip' or 'crush'

5 Crankshaft, bearings and seals

5.1 CRANKSHAFT FORM AND NOMENCLATURE

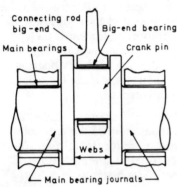

Figure 5.1 Basic crankthrow arrangement

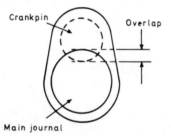

Figure 5.2 Overlap between main journal and crankpin

5.1.1 Counterbalance weights

A one-piece, as opposed to a built-up, construction is most commonly used for motor vehicle crankshafts. It consists of a series of 'crankthrows' connected together by the main bearing 'journals'. Each crankthrow is formed by a pair of 'webs', these being united by the 'crankpins' to which the big-ends of the connecting rods are coupled. As earlier mentioned in §2.4, the angular spacing of the crankthrows is related to engine balance and firing intervals.

The proportions of petrol engine crankshafts are usually such that the crankpin has a diameter of at least 0.60 of the cylinder bore dimension and a length of not less than 0.30 of the pin diameter. Web thickness of the crankthrow is generally in the region of 0.20 of the cylinder bore dimension. The main bearing journal is made larger than that of the crankpin with a diameter of up to 0.75 of the cylinder bore dimension and a length of about 0.50 of the journal diameter.

Adequate crankshaft rigidity to resist both bending and twisting is a major requirement for smooth operation. With current short-stroke engines, the proportions of the crankshaft are generally such that in themselves they contribute to greater rigidity. This results from the combination of a smaller crankthrow radius and larger bearing diameters, which permit a beneficial overlap between the main journals and the crankpins.

Since the crankshaft is subjected both to bending and torsional load reversals, it must also be designed to resist failure by fatigue. This condition may be initiated at any point where there is a concentration of stress or, in other words, a heavy loading confined to a very small area. In practice, it may occur at any abrupt change of cross-section, or from the sharp edge of an oil hole or a corner of a keyway.

When the crankshaft is rotating, centrifugal force acting upon each crankthrow and the lower part of its associated connecting rod tends to deflect the crankshaft. Since this deflection is resisted by the main bearings, their loading is correspondingly

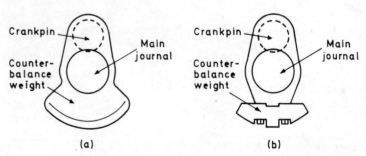

Figure 5.3 (a) Integral and (b) separately attached counterbalance weights

increased. To reduce these loads 'counterbalance' weights are either formed integrally with, or separately attached to, the crankthrow webs. The former arrangement is now most commonly used, the crankwebs being extended opposite to the crankpin and spread circumferentially.

5.2 LUBRICATION OF CRANKSHAFT BEARINGS

The object of engine lubrication is to reduce friction, heating and wear of the working parts or, more precisely, wherever interacting surfaces are in relative motion. It is, of course, accomplished by introducing a film of lubricating oil between the various bearing surfaces to keep them apart, thereby providing a condition of 'fluid' friction rather than 'dry' friction. The lubrication process exists in several forms, but that which takes place in journal bearings is chiefly that known to the lubrication engineer as 'hydrodynamic'.

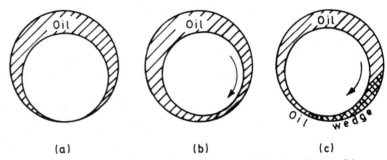

Figure 5.4 Hydrodynamic lubrication of a bearing: (a) shaft stationary, (b) rotation starts, (c) steady rotation fluid film lubrication

In hydrodynamic lubrication a converging or 'wedge-like' film of oil is established between the load-carrying surfaces of the bearing, so that they are separated by a relatively 'thick' film of oil. It is the moving surface of the bearing that acts as a pump to feed oil into the bearing clearance, which by virtue of the position adopted by the journal narrows gradually in the direction of motion, thereby generating an appreciable oil film pressure.

It should be realised that this pressure may be more than 50 times the oil supply pressure, or what is commonly termed 'engine oil pressure'. To maintain a condition of hydrodynamic lubrication it is, of course, essential that a sufficient quantity of oil is always supplied to the bearing by the engine lubrication system, as later described in §7.2.

5.2.1 Oil supply to the crankshaft bearings

Of major importance is the oil supply to the crankshaft main bearings and their dependent connecting rod bearings. Oil leaving the engine main supply gallery is therefore normally routed directly to the main bearings through communicating ducts in the crankcase webs and main bearings housings. From these ducts the oil passes through an inlet hole in the upper half-liner of each main bearing. At this point, the oil flows into a circumferential groove formed in the actual bearing surface. A proportion of the oil leaving the groove spreads laterally to lubricate and cool the main bearing, whilst the remainder is transferred to perform a similar duty in an adjacent connecting rod big-end bearing. The latter action is accomplished by means of an angular drilling which registers with the supply groove and passes through the appropriate main journal, web and crankpin.

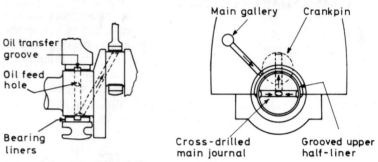

Figure 5.5 Oil supply to the crankshaft bearings

Oil grooving in a bearing liner is always a necessary evil. For some high-performance engines a part-circumferential oil supply groove may be employed to improve the load-carrying capacity of the bearings. In this case, the lower half-liner of each bearing is made grooveless, which in turn requires the main journals to be cross-drilled. Hence, either end of this drilling is always in communication with the grooved upper half-liner, thus ensuring an uninterrupted supply of oil to the big-end bearing with each full turn of the crankshaft.

5.3 CRANKSHAFT OIL-SEALING ARRANGEMENTS

Seals are fitted where each end of the crankshaft emerges from the engine structure. Their purpose is to prevent leakage of oil from the adjacent lubricated areas and also to protect the engine mechanism against the ingress of foreign matter.

The sealing devices used for this purpose fall into two categories, as follows:

(1) *'Clearance' seals:* Although no actual rubbing contact exists between the shaft and seals of this type, they nevertheless impose a hydraulic resistance against oil leakage by virtue of their return pumping action.
(2) *'Contact' seals:* Their object is to provide a controlled rubbing pressure against the shaft, thereby creating a positive sealing action.

These seals are often used in conjunction with a supplementary oil flinger ring which rotates with the crankshaft. Its action is such that centrifugal force tends to throw off any excess amounts of oil creeping along the shaft towards the main seals. The oil flinger rotates within an annular grooved housing, this in turn being provided with a drain passage for returning oil to the sump.

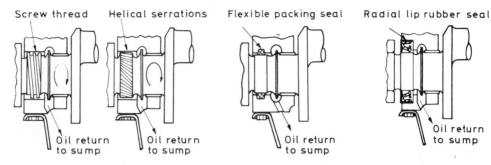

Figure 5.6 Sealing the rear end of the crankshaft

5.3.1 Sealing the rear end of the crankshaft

Here the sealing requirements are particularly exacting since the seal has to deal with the relatively large flow of oil from the adjacent rear main bearing. In many constructions the diameter of the crankshaft is substantially increased behind the rear main bearing to form an integral oil flinger, which is followed by a slightly reduced diameter sealing portion.

For clearance seals, either an oil-return thread or a band of helical serrations is usually provided on the sealing surface of the shaft, which runs within a close-clearance plain bore housing.

With the simple contact seal a flexible packing material is used, the rubbing portion of the shaft being left as a plain journal. The sealing strip is pressed into upper and lower grooves machined in both the rear wall of the crankcase and an extension of the rear main bearing cap. In modern practice, the radial lip type of contact seal made from synthetic rubber has come into widespread use. This type of seal is provided with a spring-loaded flexible sealing lip, which maintains a light rubbing contact with the shaft sealing surface. The sealing surfaces do, in fact, run with a very thin film of oil between them, the positive sealing action being generally attributed to the surface tension effect of the oil film at the exit side of the seal. A press fit must be provided for the seal in its housing and the wiping lip is positioned facing the lubricated bearing.

5.3.2 Sealing the front end of the crankshaft

The sealing requirements here are less severe, because the seal is shielded from the oil flowing from the front main bearing by the timing drive sandwiched between them, although this does not apply where an external toothed-belt timing drive is used, as described at a later stage.

The front end sealing surface is usually furnished by an inner extension of the crankshaft pulley boss. This is left as a plain journal when used in conjunction with either a threaded oil-return bush, or a radial-lip type seal, inserted in the timing cover. Alternatively, an oil-return thread may be provided on the pulley boss, which then works within a close-clearance plain bore in the timing cover. Where an oil flinger is also used, it is separately attached to the crankshaft behind the pulley boss.

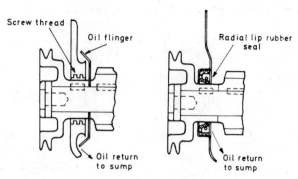

Figure 5.7 Sealing the front end of the crankshaft

5.4 CRANKSHAFT FITTINGS

These include the flywheel, timing drive and fan pulley. Also, the drive for the oil pump may be taken from the crankshaft, as mentioned in §7.3.

5.4.1 Flywheel

A one-piece construction in cast iron is generally employed for the flywheel, although in a few designs a relatively thin disc is used to mount a separate cast-iron rim. The object of the latter construction is to minimise any disturbance of the flywheel that could result from bending vibration of the crankshaft and thus promote smoother running.

Radial location of the flywheel hub is afforded by a spigot on the rear end of the crankshaft. Owing to its appreciable inertia, the flywheel is located in the rotational sense by dowel pins and clamped firmly to the rear face of the crankshaft by a ring of bolts. The rim of the flywheel provides a mounting for the starter ring gear and may also bear timing marks for checking the valve and ignition settings, relative to prescribed positions of the crankshaft.

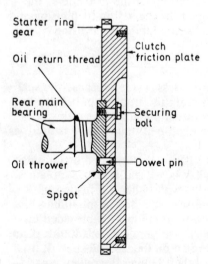

Figure 5.8 A one-piece construction flywheel

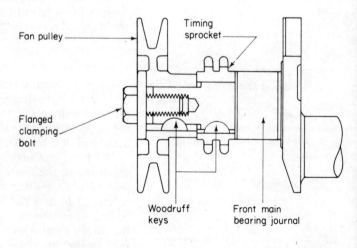

Figure 5.9 Attachment of timing wheel and pulley to nose of crankshaft

5.4.2 Timing wheel and pulley

Ahead of the front main journal of the crankshaft, a cylindrical extension is machined to accept the driving wheel of the timing drive (the various arrangements of which are described at a later stage), an oil flinger and the driving pulley for the belt-driven engine auxiliaries.

These components are made close-fitting on the shaft extension or 'nose' and are prevented from turning relative thereto by either a single parallel-face key, or a series of Woodruff keys, the latter being self-aligning by virtue of their semi-circular form.

The complete assembly of timing wheel, oil flinger and belt pulley are retained endwise on the shaft extension by means of a large setscrew which enters a hole tapped axially in the nose of the shaft and exerts its clamping load via a thick plain washer. In some engines, a driving gear for the oil pump may additionally be sandwiched between the timing wheel and pulley combination.

5.5 CRANKSHAFT MATERIALS AND MANUFACTURE

Until the early nineteen-sixties petrol engine crankshafts were traditionally forged from high-strengh, low-alloy steels, and indeed these are still used for heavy-duty applications. Since then, however, the majority of car manufacturers have favoured the use of crankshafts produced from iron castings of the 'spheroidal graphite' type or 'SG iron' as it is commonly termed.

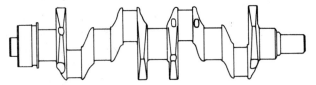

Figure 5.10 General form of a high-strength cast-iron crankshaft for a four-cylinder engine

High-strength cast irons of this type were first developed in the late nineteen-forties, both in Britain and in America, and their distinguishing feature is that the graphite structure takes the form of spheroidal nodules. This feature confers higher strength, better ductility and greater toughness than the flake graphite structure of normal grey cast iron. It is obtained by injecting a trace of magnesium into the iron melt, which causes the graphite flake to gather into little balls or, more technically, 'spheroidal nodules', that greatly strengthen the grain structure of the material.

5.5.1 Crankshaft manufacture

A forged crankshaft, like any other forged metal component, is manufactured by a process in which the metal in a more or less plastic, rather than molten, state is forced to flow into the desired shape by means of hammering, squeezing and bending. In the case of motor vehicle crankshafts produced in large quantities, the actual shaping is performed by drop-hammer forging in closed dies. The latter are upper and lower blocks of metal in each of which an impression has been formed of the crankshaft.

Figure 5.11 Crankshaft turning and grinding (Laystall)

A cast crankshaft is, in contrast, one that is manufactured by a process in which the metal in a molten state is poured into a mould and allowed to solidify. Motor vehicle crankshafts are not, however, produced in conventional sand moulds, but are cast vertically by the 'shell-moulding' process. In this technique, a thin shell-like mould of sand and synthetic resin is made by bringing these materials into contact with a heated metal pattern, the contours of which are exactly reproduced in the shell mould. Two such shells clamped together then form a complete mould.

Among the important advantages offered by the modern shell-moulding process is that castings can be produced to much closer tolerances. This in turn reduces the amount of machining required afterwards, and in the case of the crankweb faces can eliminate it altogether.

After heat treatment to remove residual stresses and to give the specified tensile strength of material, usually about 63 kg/mm^2, the crankshaft must be machined to its final dimensions. This involves principally the rough turning, finish grinding and final lapping of the main journals and crankpins. With SG iron crankshafts it is good practice that the bearing journals and pins be final-lapped with the crankshaft rotating in the same direction as it does in the engine.

5.6 CRANKSHAFT MAIN BEARINGS

In modern high-compression-ratio engines, the number of main bearings employed to support the crankshaft has tended to increase. This is because the crankshaft is subjected to greater bending loads, resulting from the higher peak gas pressures acting upon the pistons. Hence, the crankthrows must receive adequate support from adjacent bearings to minimise shaft deflections. It is therefore now customary for a main bearing journal to be used at each end of the crankshaft and between each cylinder of in-line engines, and each pair of cylinders in the more compact horizontally-opposed and V engines.

5.6.1 Form and nomenclature of the main bearings

The thin-wall main bearings are of similar form to the big-end bearings and must be rigidly supported in the crankcase so as to preserve not only the geometrical truth of their working surfaces but also their correct relative alignment with one another. These requirements must be met to avoid high localised pressures which may result in destruction of a bearing through breakdown of its oil film with consequent overheating.

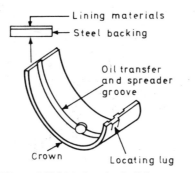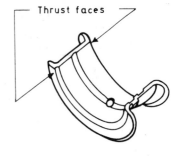

Figure 5.12 Main bearing half-liners

Similarly, the nomenclature of thin-wall main bearings follows that of the big-end bearings with the addition, of course, of an 'oil groove'. As an alternative to separate crankshaft thrust washers, one pair of the set of main bearing half-liners may be provided with integral flanged ends to serve as 'thrust faces'. Such an arrangement does, however, lack the inherent self-aligning and good heat-conducting properties of separately-mounted thrust washers.

5.6.2 Materials and location

Again, the choice of materials and the methods of location for the main bearing half-liners correspond to those of big-end practice, although another method of providing lateral location that is sometimes used for heavy-duty main bearings is the 'dowel-and-hole' arrangement. In this method the dowels are fixed securely in the crankcase bearing saddle and cap, and register with circumferentially-elongated holes in each half-liner. A disadvantage of this particular method of location is that the presence of the dowel hole disrupts the oil film in the bearing.

5.6.3 Installation of the main bearings

The fitting precautions to be observed are basically similar to those mentioned earlier for the big-end bearings. It is, of course, necessary to install the crankshaft thrust washers prior to fitting the appropriate main bearing cap, taking care that the oil grooves face the thrust surfaces on the crankshaft. The series of bearing caps are then torque tightened to their specified value, starting from the centre and working alternately towards each end. Any undue resistance to rotation of the crankshaft by hand should be checked after finally tightening each bearing cap.

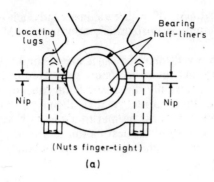

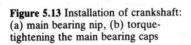

(a)

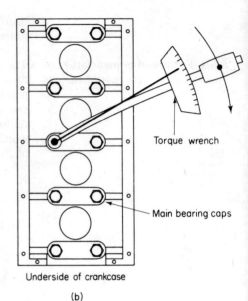

Underside of crankcase

(b)

Figure 5.13 Installation of crankshaft:
(a) main bearing nip, (b) torque-
tightening the main bearing caps

6 Overhead valve mechanisms

6.1 METHODS OF OPERATION

Before discussing overhead valve mechanisms in particular, it may be useful briefly to summarise how they became positioned in this manner.

6.1.1 Historical background

Chiefly for reasons of accessibility, the inlet and exhaust valves in very early motor vehicles were arranged in two separate rows, one on either side of the cylinders and operated from beneath by similarly-positioned camshafts. This long since obsolete arrangement of 'side valves' provided what was known as a 'T-head' engine. Its combustion characteristics were later recognised as being poor and the contrasting 'hot' and 'cold' (exhaust and inlet) sides of the engine could lead to cylinder distortion problems.

The T-head engine was then superseded by the 'L-head' type in which the inlet and exhaust valves were arranged in a single row on one side of the cylinders and again operated from beneath by a similarly-positioned camshaft. Engines with this particular arrangement of side valves underwent considerable development and for many years provided a power unit which was generally cheap to produce. It became obsolete during the mid-nineteen-fifties because its power output was somewhat limited by space restrictions on the usable size of inlet valves and by difficulties encountered in adequately cooling the exhaust valves.

An early compromise between positioning the valves alongside the cylinders and in the head above them was a combination of these two locations. This resulted in the occasional 'F-head' engine being built with 'overhead inlet' and 'side exhaust' valves, both sets of valves being operated from a single camshaft mounted in the crankcase. Each overhead inlet valve was operated by a 'push-rod-and-rocker' system. The main advantage of this type of layout was that larger inlet valves

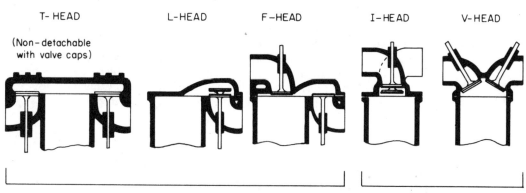

Figure 6.1 Identification of basic valve arrangements

could be used, but being heavier they also placed limitations on maximum allowable engine speed. F-head engines were relatively expensive to produce and have been obsolete since the early nineteen-sixties.

While these various developments of the side-valve engine had been applied to many touring cars, the designers of racing and sports cars had fairly early recognised that better engine performance could be more readily obtained by placing both the inlet and the exhaust valves overhead each cylinder, albeit with a certain amount of mechanical complication and less quiet operation. In varying degrees such arrangements allowed of more efficient shapes of combustion chamber and for a less tortuous route to be taken by the ingoing mixture and the outgoing exhaust gases, which otherwise slowed them down.

Furthermore, the inlet and exhaust valves could be arranged either in a single vertical or near-vertical row, or be separated into two rows and mounted at an included angle to each other. These two arrangements of in-line and inclined valves are thus said to provide 'I-head' and 'V-head' engines, respectively. Their mode of operation can either be directly from a single or a pair of cylinder-head-mounted or 'overhead' camshafts, or indirectly through a push-rod-and-rocker system acting upon one, and in some cases two, crankcase-mounted or 'side' camshafts.

The present-day requirement for high performance from a 'medium-power' engine has resulted in generally higher maximum crankshaft speeds, typically in the region of 5500–6000 rpm. This can be explained by recalling from §1.5.8 the factors governing the power output of an engine since, if neither piston displacement nor mean effective pressure can be further increased, then the only other way to raise maximum power is to permit higher crankshaft speeds. For this reason, it becomes increasingly more important to avoid erratic operation of the valves at high engine speeds. As a result, about 65% of car manufacturers now offer models with the engine valves operated directly from an overhead camshaft, rather than by the less rigid push-rod-and-rocker system.

Commonly-used abbreviations in connection with these valve layouts are as follows: SV, side valves; IOE, inlet-over-exhaust valves; OHV, overhead valves; SOHC, single overhead camshaft; DOHC, double overhead camshafts.

6.1.2 Side camshaft, push-rods and rockers

Camshaft: This serves to open the engine valves positively and to control their closing against the return action of the valve springs. In motor vehicle practice, a one-piece construction is invariably used for the shaft and its cams. Camshafts are generally produced from hardenable cast iron, which has

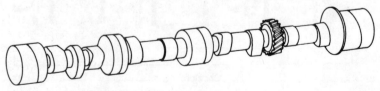

Figure 6.2 Camshaft for a four-cylinder engine

replaced the case-hardened forged steel material used formerly. The angular spacing of the integral cams is such as to impart the required motion, in correct sequence, to the inlet and exhaust valves in each cylinder. To preserve accuracy of valve motion, the camshaft must be rigid enough to resist deflection under the alternating torsional and bending loads imposed upon it by the valve operating mechanism. The camshaft journals are supported radially in the crankcase by a series of plain bearings.

Cam followers: To convert the radial motion of the cams into the reciprocating motion necessary for opening and closing the valves, requires the use of cam followers or 'tappets'. These must be used because the force exerted by a cam acts perpendicular to its contact surface and therefore does not remain in the direction of follower travel. In other words, whatever mechanism bears directly on the cam is subject to a certain amount of side thrust. Sliding-type followers are almost invariably used in conjunction with crankcase-mounted camshafts. This type of follower is termed a 'barrel' tappet and is typically produced from hardenable cast iron. However, since certain combinations of camshaft and tappet materials behave better than others, great care has to be exercised by the engine designer to ensure their compatability and avoid scuffing. Various surface treatments are also applied to the cams and tappet barrels, in order to assist the 'running-in' process of their highly-stressed contacting surfaces.

Push-rods: These are required to transmit the reciprocating motion of the cam followers to the valve rockers. Both ends of the push-rod form part of ball-and-socket joints which accommodate the angular movements of the push-rod arising from the straight line motion of the tappet barrel on the one hand and the arctuate motion of the valve rocker on the other.

Since the push-rod is part of a valve train that constitutes a vibrating system, it must combine maximum rigidity with minimum weight. Push-rods are generally produced from steel and may be of either solid or tubular construction. In comparing solid and tubular push-rods of the same strength, the latter usually offer some reduction in reciprocating weight and can also serve as an oil conduit in the valve train lubrication system.

Valve rockers: The function of these is to cause both a reversal and a magnification of the motion imparted by the cam and follower to the valve. The valve rocker, or 'rocker arm', is a short, rigid beam that oscillates about an offset pivot of either the journal bearing or the ball-and-socket type. An advantage of the latter type is that it makes the rocker inherently self-aligning, but it is necessary to introduce some means of restraining lateral movement. In cross-section the depth of the rocker arm greatly exceeds its width, since the bending loads imposed upon it are mainly within the plane of oscillation with very little side loading.

Valve rockers may be of either 'solid' or of 'hollow' construction and may be produced from steel forgings, iron castings or steel pressings. Solid rockers oscillate about a stationary hollow shaft known as the 'rocker shaft'. This is

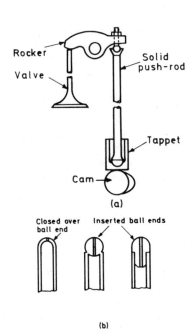

Figure 6.3 (a) Push-rod valve operation, and (b) types of push-rod

supported by a series of pedestals mounted on the top deck of the cylinder head, each pair of rockers generally being separated from the next pair by one of the pedestals. Sideways location of individual rockers is against adjacent rocker shaft pedestals, the rockers in each pair being held apart by means of a compression spring, spring clips or spacer tube. For some designs a separate bronze bushing is pressed into the rocker bearing bore, while in

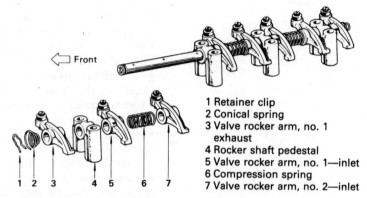

1 Retainer clip
2 Conical spring
3 Valve rocker arm, no. 1 exhaust
4 Rocker shaft pedestal
5 Valve rocker arm, no. 1—inlet
6 Compression spring
7 Valve rocker arm, no. 2—inlet

Figure 6.4 Rocker-shaft assembly for a four-cylinder engine (Toyota)

others the rocker is hardened all over with a plain bore bearing directly on the rocker shaft. A curved pad is machined on the end of the rocker where it contacts the valve stem tip, so as to allow the partly rolling and partly sliding motion that occurs between them.

Since becoming accepted practice in the USA, hollow rockers are now widely used elsewhere. Although they have been used with conventional rocker shafts, hollow rockers are now usually mounted on individual pivot posts with either hemispherical or part-cylindrical seatings. The pivot posts may be pressed and in some cases screwed into bosses on the top deck of the cylinder head.

6.1.3 Overhead camshaft and sliding followers

Camshaft: For a cylinder-head-mounted camshaft, modern practice tends to favour an increase in the number of intermediate bearings to one between each pair of cams. The

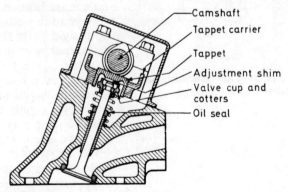

Camshaft
Tappet carrier
Tappet
Adjustment shim
Valve cup and cotters
Oil seal

Figure 6.5 Overhead camshaft and sliding followers

bearing installations for overhead camshafts are currently similar to those of crankcase-mounted camshafts. An exception is where the bearings are supported in pedestals with detachable caps, which permit the camshaft to be inserted from above the engine. Also, this particular arrangement enables a reduction to be made in the diameter of the journal bearings, since the cams no longer have to be assembled through them.

Cam followers: The use of direct-acting sliding followers with a cylinder-head-mounted camshaft demands relatively large diameter tappets. This increase in their base area is dictated by the absence of a multiplying leverage (otherwise provided by rocker arms) in the valve train, so that larger-size cams have to be used to give the desired amount of valve lift. Since an overhead camshaft is in even closer proximity to the valves than it was in the obsolete side-valve layout, the tappets must necessarily be both hollow to fit over the valve springs and short in length. For this reason they are aptly termed 'inverted-bucket' tappets.

6.1.4 Overhead camshaft and pivoting followers

Camshaft: Similar remarks apply as in §6.1.3 regarding the number and installation of the camshaft bearings.

Cam followers: These are known as the 'pivoting' type because they swing between the camshaft and the valve tips and act as either a simple (straight) lever arm or 'bell-crank' (angled) lever.

Modern applications of the former type generally feature a forged steel lever arm, or 'finger rocker' as it is often called, which is self-aligning on a ball-and-socket pivot mounting, although journal bearing pivots have been used in the past. To restrain lateral movement of the ball-and-socket-mounted finger rocker, the valve stem tip may be recessed into the curved contact pad on the end of the arm.

The 'bell-crank' type of pivoting rocker is supported on journal bearings from a conventional rocker shaft. This is mounted immediately above the camshaft, so that the downwards extending lever arms of the bell-cranks bear on the cams and the outwards extending arms contact the valve tips.

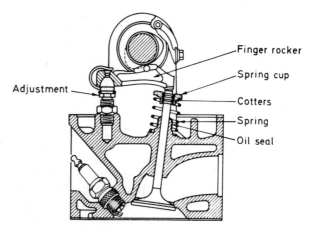

Figure 6.6 Overhead camshaft and pivoting followers

6.1.5 Advantages and disadvantages

Side camshaft, push-rods and rockers: The advantages of this method of valve operation are generally that the timing-drive to the camshaft is uncomplicated and therefore less expensive to produce, and that tappet clearance adjustment can be conveniently performed. Disadvantages are concerned with its greater tendency towards vibration, which is due to flexing of the push-rods and rockers and creates false motions of the valves. There may also be the need for not-infrequent tappet clearance adjustment because of the number of points subject to wear.

Overhead camshaft and sliding followers: For this method of valve operation the advantages are chiefly its lower inertia and greater rigidity in the interests of high-speed operation and a reduced requirement for tappet clearance adjustment. Its disadvantages include the need for a more elaborate and therefore expensive timing-drive to the camshaft(s); the engine in general tends to be more difficult to service, although tappet clearance adjustment in particular need not be laborious in modern versions, and in any event should seldom be required; and the engine height is increased.

Overhead camshaft and pivoting followers: Here the advantages are that it can provide a further reduction in inertia over that of sliding followers, together with some magnification of the motion imparted by the cam to its valve. It is also simpler to incorporate a ready means of tappet clearance adjustment. Disadvantages are generally that finger-rockers do not possess the same degree of rigidity as sliding followers; there must inevitably be some side thrust transmitted to the valves and effective lubrication can be more difficult to attain for minimum wear.

6.2 VALVE CLEARANCE ADJUSTMENTS

The design of an engine must take into account the effects of thermal expansion and contraction of the valve train components, relative to that of the engine structure. Since these effects could result in the valves being held off their seats when they should be firmly closed, a small operating clearance has to be introduced into the reciprocating parts of the valve train. This is generally termed the 'valve or tappet clearance', and provision for either its manual or automatic adjustment is incorporated in all engines. The latter form of adjustment will be dealt with at a later stage.

6.2.1 Methods of manual adjustment

For engines with push-rod overhead valves and solid rockers, valve clearance is usually adjusted at the rockers by means of a locking nut and hardened screw, the ball end of which registers in the push-rod socket. The clearance may be checked by inserting a feeler gauge blade of appropriate thickness between the rocker pad and the valve stem tip, a useful practical hint being not to slacken completely the locking nut while turning the adjusting screw.

Where manual adjustment of the valve clearance is employed for push-rod overhead valves and hollow rockers, a screw-threaded portion at the upper end of the pivot post receives a

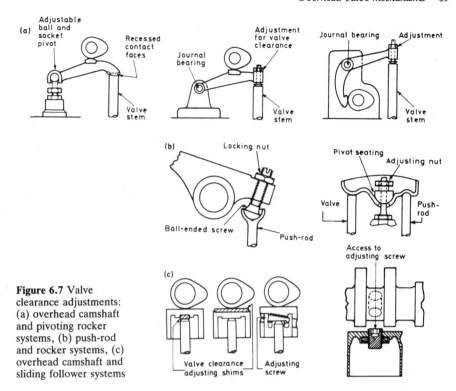

Figure 6.7 Valve clearance adjustments: (a) overhead camshaft and pivoting rocker systems, (b) push-rod and rocker systems, (c) overhead camshaft and sliding follower systems

self-locking nut, against the underside of which the pivot seating abuts. Turning the self-locking nut thus allows the valve clearance to be set by virtue of either lowering or raising the rocker pivot. In the case of an overhead camshaft and inverted bucket tappets, the adjustment can be a time-consuming operation in those installations where valve clearance is set by the selective assembly of graduated-thickness shims, which are inserted between the valve stem tip and the underside of the tappet head. Other less inaccessible means of adjustment have, however, been introduced in recent years for inverted-bucket tappets.

In one system discs of appropriate thicknesses are located in recesses in the tappet heads and are directly contacted by the cams, so that if a valve is held depressed after being operated by its cam the disc may easily be changed. Another system utilises the wedging action of a taper-faced adjuster-cum-locking screw, which is located transversely in the tappet and forms an abutment for the valve tip. Yet another and particularly ingenious system incorporates a self-locking screw in the head of the tappet. Access to this screw, which bears directly against the valve stem tip, is gained through a divided cam and cross-drilling in the camshaft.

For manual adjustment of valve clearance with an overhead camshaft and self-aligning finger-rocker, the ball pivot is usually screwed into a boss in the top deck of the cylinder head, so that raising or lowering the ball-pin alters the valve clearance accordingly. In the case of journal bearing pivots, a screw-type adjustment for valve clearance is generally provided at the valve tip end of the rocker. The same applies to the bell-crank type of pivoting rocker. An interesting point to note in connection with these various methods of valve clearance adjustment is that those where the screw adjuster remains stationary serve to reduce the inertia of the valve train.

6.3 METHODS OF VALVE-SPRING RETENTION

In modern engines the different methods of valve-spring retention are related to the varying extents that rotation of the valves is encouraged. Gradual rotation of the valves during engine operation is usually recognised as being beneficial to valve life. This is because it promotes a more uniform temperature distribution around the valve seat, as well as improving heat transfer from the valve by virtue of more effective removal of seat deposits. The valves are generally

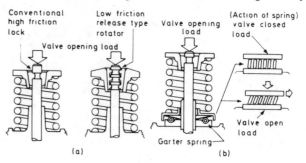

Figure 6.8 Methods of valve spring retention

credited with an inherent tendency towards rotation, which is variously attributed to engine vibration and the slight winding-up of the valve springs as they are compressed when the valve is opened. Several devices of American origin may be employed, however, to impart either a 'non-positive' or a 'positive' rotational movement to the valves.

6.3.1 High-friction-lock retention

This was once the conventional method of retaining a valve spring, the moving end of which encircles a spigot washer with a central conical recess. Seating in the latter is a pair of split-type 'wedge collets' which in turn engage a shallow groove machined near the top of the valve stem. The pressure exerted by the valve spring therefore holds the collets tightly in place, yet they can readily be removed (albeit with a little persuasion from a soft-faced mallet in some cases!) by depressing the spring while holding the valve closed. This method of valve spring retention makes no pretence at encouraging rotation of the valves during engine operation.

6.3.2 Release-type rotators

This method of valve spring retention confers a 'non-positive' rotary movement to the valves by merely permitting them to rotate relatively freely under the influence of engine vibrations. Such action is made possible by dispensing with the radial clamping exerted by the high-friction-lock method of retention and transmitting the spring load to the valve stem through a multiple-thrust collar arrangement, which is provided by abutting collets. This is in contrast to the gap that must necessarily exist between the sides of wedge collets.

6.3.3 Positive-type rotators

In this method of valve spring retention, valve rotators of the 'positive' type are incorporated as intermittent motion devices. A modern version consists of an enclosed garter-spring, which is sandwiched between a thrust washer and the lower valve spring washer, there being the normal high-friction-lock against the upper end of the valve stem. During operation the increase in

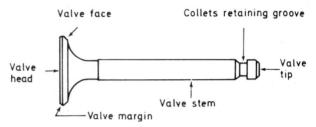

Figure 6.9 Terminology of the poppet valve

valve spring load that accompanies valve opening causes the coils of the garter spring to lean over. In so doing they act as friction sprags and transmit a torque to the valve spring and its upper retaining washer, which in turn imparts a small rotational movement to the valve itself.

6.4 VALVES AND SPRINGS

It has long been established practice to employ what are termed 'poppet' valves, which comprise a disc-shaped head with a conical seating and a stem that acts as a guiding surface. Their main advantages over other possible alternative valve forms are as follows:

(1) They are self-centring as they close onto the cylinder head seating.
(2) They possess freedom to rotate to a new position.
(3) It is relatively easy to restore their sealing efficiency in service.

For mechanical strength and to assist gas flow, the valve stem is blended into the head portion to form a neck of fairly generous radius, but a certain amount of relative flexibility between them is generally permitted. Under the influence of cylinder pressure, the valve head may therefore better conform to its seating should there be any distortion present. The valve stem is provided with a few hundredths of a millimetre working clearance in its guide, this usually being increased slightly for the hotter exhaust valves to allow for their greater expansion.

It is usual for the head diameter of the exhaust valve to be made less than that of the inlet valve, since cylinder gases may be more easily evacuated at exhaust pressure than admitted at induction depression—truly it is said that 'nature abhors a vacuum'! Another reason for favouring a smaller-diameter exhaust valve is generally to reduce its thermal loading by virtue of the shorter path of heat flow.

The conical facing of the valve makes an angle of either 45° or 30° with the plane of the head. Although the former angle provides a higher seating pressure for a given valve spring load,

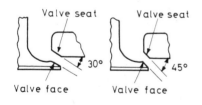

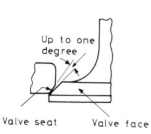

Figure 6.10 Valve seating angles

the latter angle permits a greater gas flow for a given valve lift. Hence in some engines the face angles of the exhaust and the inlet valves may be 45° and 30° respectively. To improve further their seating conditions, the exhaust valves may be installed with a differential face-to-seat angle of up to 1°. The effect of this is initially to concentrate the seat bedding towards the larger-diameter and combustion side of the valve face, so that any subsequent dishing of the valve head then tends to centralise the seat bedding.

In practice, the face angle of the valve may therefore either match, or have a slight positive interference, with the seat angle. Also, the actual bedding area should be neither too narrow, which would hinder heat dissipation, nor too wide, which would reduce seating pressure and be less effective in breaking up deposits.

6.4.1 Valve materials

The valves must endure a particularly arduous existence, since they are repeatedly subjected to severe mechanical and thermal loading. Exhaust valves have to withstand mean operating temperatures that may approach 900°C, so they must be produced from special high-alloy steels. Modern practice generally favours the stainless austenitic alloys which contain a fairly high proportion of chromium. These are employed not only for their resistance to corrosion attack, but also because they maintain both their strength and hardness at high temperatures.

Since the thermal loading on the inlet valves is less severe, they are generally produced from low-alloy steels containing a lesser proportion of chromium. The durability of the inlet valves may sometimes be enhanced by the application of an aluminium coating to their heads. Both the inlet and the exhaust valves may have their stems chromium-plated to resist wear.

6.4.2 Valve springs and materials

In conventional practice, the valve springs are of helical, or coil, type consisting of wire wound in the form of a helix. The valve closing load is conveyed axially along the spring, which stresses the material principally in torsion. For the valve open and the valve closed conditions, the ratio of spring loads is usually in the region of 2:1. Since the valve spring is compressed between parallel abutments, its end coils are ground flat and square with the spring axis. The coil ends are also diametrically opposed, so as to minimise an inherent tendency towards bowing of the spring as it is being compressed. Because they are subjected to severe service, valve springs are produced from high-duty materials, these generally being either hard-drawn carbon steel or chrome–vanadium steel. To reduce stress concentration in the spring wire and thus make it less liable to fail by fatigue, the valve spring may be 'shot-peened'. In this process the wire surface is bombarded at high velocity with metal shot, which induces a residual compressive stress that discourages crack propagation in the material.

7 Engine lubrication system

7.1 ENGINE OIL FILTRATION

Apart from corrosive wear arising from the acidic products of combustion, the other major form of engine wear is that caused by the presence of abrasive particles in the lubricating oil. This is especially true of those particles of a certain critical size that can span the oil film in the working clearances of the engine parts. Although manufacturers take immense care to build 'clean' engines, traces of casting sand and machining swarf are not unknown sources of abrasive particles. The 'running-in' process can also be responsible for generating wear particles. During normal running any soot particles contained in the cylinder blow-by gases and hardened by high temperatures can enter the crankcase oil as abrasive particles.

Since the engine lubrication system can never be any better than its oil, it is therefore necessary to include some means of filtering the oil circulating in the system, so as to remove the abrasive particles before they reach the sensitive parts of the engine.

7.1.1 Full-flow filtration

An absorbent filter of the type described in §7.5.2 may be arranged to arrest abrasive particles from either a small proportion, or practically all, of the oil being delivered by the pump to the principal working parts of the engine. Modern practice favours the 'full-flow' system in which all the oil is filtered continuously, except that discharged from the main pressure-relief valve. Exceptionally this valve may be situated on the outlet side of the filter for better protection from contaminants, in which case the filter takes the full delivery

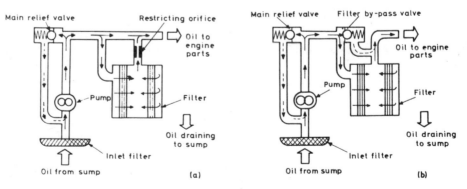

Figure 7.1 Comparison of (a) partial, and (b) full-flow systems of oil filtration

from the pump. The full-flow filter is additionally equipped with its own internal relief valve. Its purpose is to by-pass an emergency supply of unfiltered oil to the engine bearings, should the filter element become choked through neglect.

7.1.2 Partial-flow filtration

In this system, which can now be regarded as obsolete, an absorbent type of filter is arranged in the engine lubrication circuit such that 5–10% of the oil delivered by the pump is diverted through it and returned to the sump. Hence, the supply of oil to the bearings is not directly filtered or, in other words, only a small proportion of the circulating oil is filtered. To maintain an adequate supply of oil to the engine bearings the flow through the filter is further limited by a restricting orifice. This is usually situated on the outlet side of the filter, so that it is less likely to become obstructed. An internal relief valve is not required for a partial-flow filter, since if the element becomes choked the oil that would otherwise pass through it simply continues with the main oil flow to the bearings.

It is possible to employ a very fine degree of filtration with the partial-flow system, on account of the slower rate at which oil is passing through the filter. However, the widespread adoption of the full-flow system in which practically all the oil delivered by the pump is subject to immediate filtration, seems to confirm that it performs the filtration process with more certainty. Indeed, it is now recognised in the field of hydraulic engineering that it is statistically impossible to clean the whole fluid content of a system by partial-flow filtration.

7.2 ENGINE LUBRICATION SYSTEMS

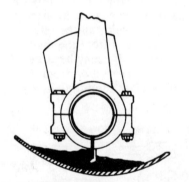

Figure 7.2 Early splash system of engine lubrication

A 'splash' type of lubrication system was used for the engines of early motor vehicles. Lubrication of the working parts was initiated simply by allowing the connecting rod big-ends to dip into a trough of oil in the crankcase, each time they passed through their BDC positions. To improve the lubrication of their bearings, the big-ends were often provided with oil scoops. Surplus oil that was flung upwards into the crankcase was then collected in galleries from whence it gravitated to the main bearings, camshaft bearings and timing drive. The cylinder walls were similarly lubricated by the oil that was splashed onto them.

Following the purely 'splash' system of engine lubrication came the combined 'pressure-and-splash' or 'semi-forced' system. The essential features of this system were that an engine-driven oil pump delivered oil only to the main bearings, the big-end bearings and other working parts being lubricated in the manner of the purely 'splash' system. An attraction of this system was that no oilways had to be drilled in the crankshaft, as necessary with the 'fully-forced' system described below.

The eventual necessity to provide the later high-speed engine with a 'fully-forced' lubrication system, arose from the ever-increasing loads and speeds at which the bearings were expected to operate. This meant that the bearings not only had to be lubricated, but also cooled, by the oil circulating through them. We therefore arrive at the long-established 'fully-forced' lubrication system, in which oil under pressure is supplied to the crankshaft main bearings, connecting rod big-end bearings and camshaft bearings. The first fully-forced lubrication system for motor vehicle engines is generally attributed, like so many features of automobile practice, to the early work of Dr F Lanchester and found in the cars that once bore his name. Even so, the basic idea of forced-feed lubrication to all bearings was

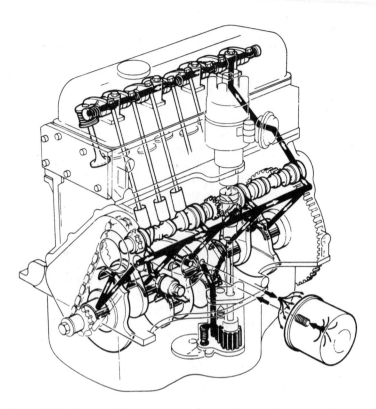

Figure 7.3 Fully-forced lubrication system for in-line cylinder engine (Renault)

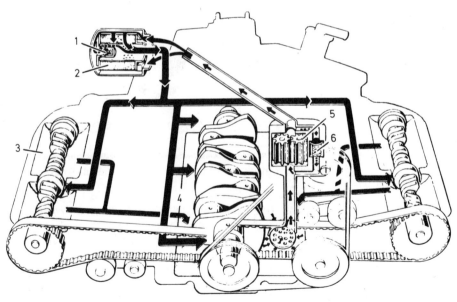

Figure 7.4 Fully-forced lubrication system for horizontally-opposed cylinder engine (Alfa Romeo)

1 By-pass valve
2 Oil filter element
3 Camshaft housing
4 Oil sump
5 Gear pump
6 Pressure relief valve

applied as long ago as 1890 in the Bellis and Morcom steam engine.

With the fully-forced oil circulation system, a major contribution to cooling of the bearings is therefore made by passing a large quantity of oil through them, in excess of the amount needed for lubrication alone.

7.2.1 The oil circuit   This consists of a network of oilways embodied in both the structure and the mechanism of the engine. Extensive use was once made of separate pipes and connecting T-pieces to convey the oil to the various working parts, but since fracturing and loosening of such piping was always a possible source of unreliability, drilled pressure conduits have long since been preferred.

In operation, the engine-driven pump delivers oil upwards through a duct in its body, whence it passes into a full-flow filter mounted on the outside of the crankcase. From the filter the oil is directed to either a single or a pair of main oil 'galleries' which extend practically the full length of the crankcase at about the level of the cylinder skirts. A single main gallery is incorporated in the crankcase sidewall of in-line cylinder engines, while a pair of centrally-disposed galleries may be required in the crankcase of some V and horizontally-opposed cylinder engines. The purpose of the main gallery is to distribute the oil to the principal working parts of the engine.

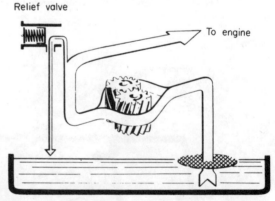

Figure 7.5 Basic action of the engine oil pump and relief valve (Alfa Romeo)

At this stage it is only necessary to gain an overall picture of the engine lubrication system, as found in typical modern practice. However, the oil supply to the crankshaft main and big-end bearings was explained in some detail in §5.2.1.

7.3 OIL PUMPS   In modern practice 'positive displacement' pumps of the rotary type are used to deliver oil and develop pressure in the engine lubrication system. The term 'positive displacement' is used in hydraulic engineering to describe any pump that displaces a definite volume of fluid with each revolution of its drive shaft, regardless of pressure conditions in the circuit it supplies. In other words, the pressure developed by the oil pump depends entirely upon the resistance offered to the flow of oil in the rest of the lubrication circuit. The principal advantage of using this type of pump is that it is compact in design. Because it is not self-priming, the oil pump with its inlet arrangements is usually bolted to a boss on the lower face of the crankcase, so that it is partially submerged in the oil sump but supported independently of it.

7.3.1 External gear pump

This comprises simply an elongated housing and end cover, which enclose a pair of either spur or, for quieter operation, helical gears. A bearing bore in the housing receives the pump drive shaft, rigidly secured to which is the driving gear of the pair. These parts may be either keyed together or machined as

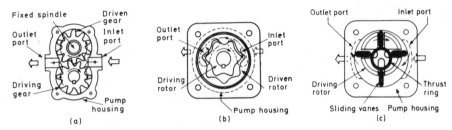

Figure 7.6 Types of engine oil pump: (a) external gear, (b) internal gear, (c) sliding-vane

an integral pair. A mounting for the driven or 'follower' gear is provided by a spindle supported from the pump housing. The gears generally have a coarse pitch tooth form, so that the least width of gear is required to pump a given amount of oil.

In operation, the oil entering the pump through the inlet port becomes trapped between the teeth of the contra-rotating gears and the surrounding wall of the pumping chamber. The oil is thus carried around the periphery of the gears (not between them) and then discharged through an opposite outlet port. At this point the action of the intermeshing gear teeth prevents the oil from returning to the inlet side of the pump.

7.3.2 Internal gear pump

This houses a smaller driving gear eccentrically-mounted within a larger driven gear. The externally-toothed driving gear is rigidly secured to the pump drive shaft, while the internally-toothed driven gear runs directly in the cylindrical cavity of the pumping chamber. Since the tooth form of these gears more closely resembles that of a lobe, the driving and driven gears are generally termed the inner and outer 'rotors' of the pump respectively. The number of lobes on the inner rotor is one less than on the outer and so determines the displacement of the pump.

In operation, the oil enters the pump through a kidney-shaped inlet port, as the lobes on the rotors are moving out of mesh. Continued turning of the rotors serves to transfer the oil to a similar kidney-shaped port on the outlet side of the pump, the oil being discharged as the lobes of the rotors move into mesh. The advantages usually claimed for this type of pump are

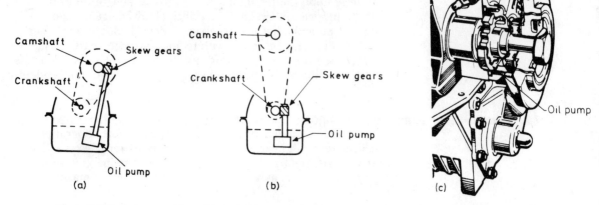

Figure 7.7 Typical pump drives: (a) from side camshaft, (b) from crankshaft, (c) encircling crankshaft (BL)

that it is still more compact than the external variety and, because there are fewer teeth in mesh for each revolution, it can be quieter running.

More recently, there have been several examples of 'crescent-type' internal gear pumps, encircling and driven directly from the nose of the crankshaft. In these cases the gear teeth are of conventional form but are separated over their non-meshing region by a crescent-shaped partition which acts as a seal between the inlet and outlet ports. Between entering and leaving the pump the oil is thus divided into two streams.

7.3.3 Sliding-vane pump   This takes the form of a driven rotor that is eccentrically-mounted within a circular housing. The rotor is slotted and equipped with vanes equally spaced around its periphery, these vanes being free to slide radially within the slots. A pair of centring thrust rings maintain the vanes in close clearance with their track in the housing.

In operation, the vanes are, of course, pressed outwards against the track by their centrifugal action. The oil inlet port is sited where the vanes begin to move away from a point of minimum volume between the eccentric rotor and its housing. After the vanes have passed the point of maximum volume and are again approaching a point of minimum volume, the oil is discharged through an outlet port. This type of pump has the characteristic of giving a continuous rather than a pulsating oil flow, the latter usually being associated with gear-type pumps.

7.4 PRESSURE RELIEF VALVES   The purpose of the pressure relief valve is to act both as a pressure regulator and as a safety device in the engine lubrication system.

A means of pressure regulation is required, since at higher engine speeds the increase in the quantity of oil delivered by the pump is not matched by a corresponding increase in the amount of oil that is able to flow through the bearings. This is because of centrifugal action on the oil already in the ducts of the crankshaft main journals, which imposes an increasing resistance to further oil flowing into them en route to the connecting rod big-end bearings.

Under low-temperature starting conditions any increase in oil viscosity (a term explained in §7.7.1) or reduction in bearing oil film clearances, or both, imposes an appreciable resistance to the oil flowing in the engine lubrication system. The pressure relief valve must, therefore, act as a safety device to prevent the oil pressure building up to a dangerous level. Otherwise, at worst the pump driving mechanism could be overloaded to its detriment, and at least the excessive pressure drop across the oil filter would cause its own internal pressure relief valve to open and bypass unfiltered oil to the engine bearings.

7.4.1 Types of pressure relief valve

The pressure relief valve consists of either a 'ball', a 'poppet' or a 'plate' valve that is spring-loaded on to an orifice seating and opened by oil pressure.

A disadvantage of the simple ball valve, which has complete freedom to turn in any direction, is that random bedding marks on its working surface can form leakage paths across the valve seating. Leakage arising from this source can be avoided with a

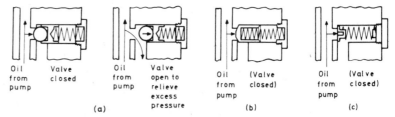

Oil from pump   Valve closed   Oil from pump   Valve open to relieve excess pressure   Oil from pump   (Valve closed)   Oil from pump   (Valve closed)

(a)   (b)   (c)

Figure 7.8 Pressure relief valves for engine lubrication systems: (a) ball type, (b) poppet type, (c) plate type

guided poppet valve, because this can turn only about its lift axis, so that the bedding condition between the valve and its seating is better preserved.

The poppet valve takes the form of a plunger with either a conical or a hemispherical sealing surface. To minimise friction of the valve sliding in its bore, the plunger guide surface may incorporate annular grooves for oil retention.

A plate-type valve possesses the least inertia and is comprised simply of a thin steel disc. Its edge is usually relieved in several places to prevent cross-binding of the valve in its guide bore.

7.4.2 Pressure relief valve setting

The actual setting of the pressure relief valve is generally related to the rate of oil flow required for adequate cooling of the bearings. In practice, it is typically set to open when the delivery pressure exceeds $280 \, \text{kN/m}^2$, although higher maximum opening pressures of up to about $420 \, \text{kN/m}^2$ are sometimes specified. The maximum delivery pressure does not, as explained in §5.2, bear any relationship to the pressures generated in the bearing oil films.

7.5 TYPES OF OIL FILTER

Various methods of oil filtration may be encountered in the lubrication system of the motor vehicle engine and they can be conveniently summarised as follows:

(1) *Surface filtration:* For this a 'coarse' type of filter is used, which simply comprises a wire gauze screen such as is fitted over the inlet to the oil pump. Its filtering action is in the manner of a sieve and it is, of course, cleanable upon removal of the sump.

(2) *Depth filtration:* This implies a 'fine' type of filter that typically comprises a specially processed paper element which is housed within a metal cannister made detachable from the engine crankcase. Because of its low structural rigidity, the paper element is provided with a supporting cylindrical metal screen. It also generally assumes a pleated form to present the largest surface area of filtering media for a given size of element. A depth filtering action is obtained as contaminants are caught, and retained, as the oil passes through from the outside to the inside of the porous material.

In modern practice, the pore size of the paper-type element may be controlled to arrest particles down to within the region of 5 μm in size, a micrometre (μm) being equal to one thousandth of a millimetre. (For purpose of comparison, an ordinary grain of table salt measures 100 μm.) As

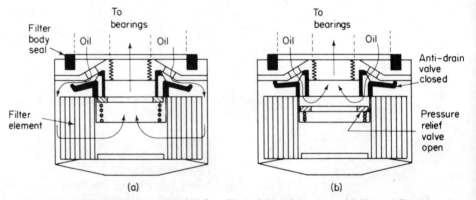

Figure 7.9 A modern full-flow filter of the spin-on type. (a) Normal filtering action. (b) Filter choked and oil by-passed over non-contaminated surfaces

deposits accumulate, the absorbent filter offers increasing resistance to the passage of oil and must therefore be replaced at appropriate servicing intervals, usually every 5000–6000 miles (8000–10 000 km). Current (full-flow) depth filters are generally sealed and of the 'spin-on' type screw fitting, so that service replacement of the element itself is no longer possible, the entire filter being fully disposable.

Another feature that may be found in the modern full-flow filter is the positioning of the element relief valve at the inlet end. The purpose of this is to ensure that any oil bypassed by the relief valve does not pass over contaminated surfaces within the filter casing.

(3) *Centrifugal filtration:* The centrifugal type of filter acts by subjecting the oil entering it to a high centrifugal force. In so doing it separates the heavier contaminants from the

lighter oil, depositing them in sludge traps formed in the perimeter of a rotating filter bowl. The latter may be mounted on the nose of the crankshaft and is of large capacity.

(4) *Magnetic filtration:* In practical application this generally comprises a plug that replaces the ordinary sump drain plug and incorporates a powerful permanent magnet. It is thus capable of attracting and retaining ferrous metal particles only, although other contaminants may also be retained by 'agglomeration', or collecting in a mass.

7.6 ENGINE OIL RETENTION

If leakage of oil from the engine is to be prevented, it must be provided with oil sealing devices of both the 'static' and the 'dynamic' varieties.

7.6.1 Static seals

These are required because to permit assembly of the working parts into the engine, its main structure must necessarily be fitted with detachable oil-tight covers. Although the mating faces of these parts are produced flat and parallel, in practice there are always minute surface irregularities present. If then a metal-to-metal closure was relied upon, these surface irregularities would offer potential areas of oil leakage. Suitably shaped static or 'gasket' types seals are therefore inserted between the various joint faces to prevent oil leakage. The gaskets are manufactured from fibre composition, cork with rubber binders and synthetic rubber materials, so that when compressed by bolt loading they tend to fill any leakage paths.

Typical static sealing applications are: the timing cover, for which a fibre composition gasket may be used; the oil sump that usually requires a cork-based gasket to accommodate the

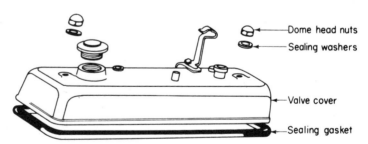

Figure 7.10 A typical static sealing application (Toyota)

relative expansion and contraction between itself and the crankcase; and the valve rocker cover where a synthetic rubber seal is generally appropriate in allowing effective sealing for minimum screw tightness. A more recent development for static sealing applications is the use of an adhesive sealant instead of a separate gasket material. For example, the camshaft cover is sealed to the cylinder head of the current British Leyland O series engine by a room-temperature vulcanising (RTV) sealant, which is claimed to increase reliability by overcoming the risk of gasket distortion.

7.6.2 Dynamic seals   No further comment is necessary here on dynamic seals, which are required where each end of the crankshaft emerges from the crankcase, since these were discussed under the heading of crankshaft oil sealing arrangements in §5.3.

7.7 ENGINE LUBRICATING OIL

Lubricants can be produced from animal, vegetable and mineral oils, but it is the latter refined from petroleum and treated with additives that are suitable for the lubrication of motor vehicle engines.

7.7.1 Functions and properties of engine oil   The importance of using the correct type and grade of engine oil will be better appreciated if first we consider the essential functions it must perform. These may be summarised as follows:

(1) *Lubrication:* The 'viscosity' of the oil is its single most important property, since it represents the internal friction in the film of oil and is caused by the resistance to shearing motion of the oil molecules. More simply, it may be regarded as the 'resistance to flow' of an oil and it is this property that makes it possible to separate the bearing surfaces in relative motion. To cater for the more difficult conditions of lubrication, such as may arise between the highly stressed cams and tappets, extreme pressure, or 'EP' additives, can be incorporated in the oil. These improve its 'oiliness' by forming a chemical film between the rubbing surfaces.

(2) *Sealing:* To assist in sealing against the leakage or blow-by of the cylinder gases, the engine oil must be physically capable of filling the minute leakage paths and surface irregularities of the cylinder bores, pistons and rings. It must therefore possess adequate 'viscosity stability' because of the high-temperature conditions to which it is exposed.

(3) *Cooling:* The conditions under which the oil functions in a cooling capacity generally tend to promote 'oxidation'. This is because the oil is exposed not only to heating—the temperature in the top compression ring groove of the piston may reach 250°C, while even that in the crankcase is typically around 125°C—and agitation, but also to the effects of contact with metals and contaminants. Therefore, to avoid the formation of gummy deposits and piston ring sticking, engine oils are treated with 'anti-oxidant' additives.

(4) *Cleaning:* The lubricating oil must also contribute to the internal cleanliness of the engine and is therefore treated with so-called 'detergent additives'. These can be more accurately described as 'dispersent additives' because they retain insoluble contaminants in fine suspension, which are removed when the engine oil is changed. Similarly, the corrosive action of acidic products that can contaminate the oil may be neutralised by treating it with 'anti-corrosion' additives.

7.7.2 Classification and grades of oil   In the mid-nineteen-twenties the American Society of Automotive Engineers (SAE) recommended an arbitrary classification system for engine oils in terms of viscosity only.

This system, which has been revised from time to time, has long since been internationally recognised. It established a numerical relationship, defined by 'SAE numbers', between the viscosities of oils, and specifies minimum and maximum viscosities at stated temperature for an oil to meet a particular grade. On this basis, SAE 20, 30, 40 and 50 grades of oil are intended for normal temperature conditions and are restricted by a minimum viscosity at 98.9°C. Suffix W grades of oil in the range SAE 5W, 10W and 20W are intended primarily for low-temperature conditions and are restricted by a maximum viscosity at − 17.8°C.

In common with other liquids, an increase in the temperature of engine oil will cause a decrease in its viscosity. As this effect could be quite pronounced with engine oils, which are required to function over a very wide temperature range, they may be treated with further chemical additives known as viscosity index or 'VI improvers'. The term 'viscosity index' expresses the degree of variation of viscosity with temperature, so the higher its value the less will be the change in viscosity of the oil with temperature.

The development of special viscosity index improvers resulted in the introduction of 'multigrade' oils in the early nineteen-fifties. These oils are formulated to meet the viscosity requirements of more than one SAE grade and are therefore classified in double SAE Numbers, such as 20W/50. The advantages originally claimed for multigrade oils included easy starting of the engine from cold with minimum oil consumption when hot.

With regard to cold starting, there is a minimum cranking speed below which an engine will not start, this speed being higher for a diesel engine than it is for a petrol engine. A manufacturer therefore determines what is termed the 'viscosity requirement' of an engine, this being the highest value of the sump oil viscosity above which the cranking speed does not exceed the minimum starting speed. Such tests are, of course, related to the cold starting temperatures likely to be encountered in those parts of the world where the car is sold.

7.8 OIL SUMP CAPACITY

The amount of oil carried in the sump necessarily represents a compromise, since too great a capacity would hinder warm-up of the engine, while too small a capacity would hasten its contamination. In practice, sump oil capacities generally range from about 3.5 to 7.0 litres. A simple dipstick is all that is generally provided to indicate the minimum and maximum acceptable oil levels.

7.8.1 Effects of incorrect oil level

(1) *Oil level too high:* If the rotating crankshaft approaches too close to the surface of the oil, then the 'windage' it creates can produce foaming of the oil, so that air enters the engine lubrication system and causes inconsistent oil pressure. Should the big-ends of the connecting rods actually dip into the oil, then both its temperature and consumption will increase. The latter because the amount of oil splashed on to the cylinder walls could exceed the controlling ability of the

piston rings, so that some of it enters the combustion chambers and is burnt.

As an interesting aside, it was once reported that in oil consumption tests on a fleet of Paris taxis, the average oil consumption was greater for those vehicles where the operator added oil more often and nearer to the dipstick MAX mark, than in others where the operator added oil less often and below the MAX mark.

(2) *Oil level too low:* This is potentially a more dangerous state of affairs, since if the oil level falls completely below the intake of the oil pump, then obviously no oil will reach the engine working parts, with disastrous results. Even if the oil level has not fallen seriously below the MIN mark, engine oil pressure can still become erratic with any surging of the oil, due to the changing motion of the vehicle. Again this is because the pump will not remain primed with oil and air is momentarily drawn into it, a situation that will be further aggravated by the build-up of foam in the oil.

7.9 REASONS FOR CHANGING THE OIL

There are two main causes of engine oil deterioration in service, these being oxidation and contamination, as already mentioned. Suffice it here to summarise matters as follows.

Oxidation: —occurs in the hottest parts of the engine
—thickens the oil and makes cold starting difficult
—reduces oil flow until the engine has warmed up
—produces lacquer deposits that can obstruct fine clearances
—corrodes the internal surfaces of the engine.

Contaminants: —formed of soot and water from combustion
—composed of lead compounds from fuel additives
—abrasive particles admitted through air intake
—tend to clog the oil filtering arrangements
—form lacquers independent of those from oxidation.

7.9.1 Oil-change intervals

The reason for periodically changing the engine oil is therefore not because it simply 'wears out', but as a result of the oxidation and contamination accompanying the depletion of some (not all) of its additives. These are, of course, gradually destroyed in the process of doing their intended job, which in modern oils they do exceedingly well. Either the depletion of additives, or the actual engine condition, forms the basis upon which the vehicle manufacturer will specify the oil change interval in service.

Current practice is typically that of changing the engine oil (and renewing the oil filter) either every 6000 miles (10 000 km) or six months, whichever comes first. More frequent oil changes may sometimes be specified when the car is subjected, for example, to a fair amount of 'stop–start' driving.

8 Petrol engine fuel system

8.1 GENERAL LAYOUT OF SUPPLY SYSTEM

The purpose of the fuel supply system is to store, transfer and filter the petrol required either by the mixing chamber of the carburettor, or the reservoir of a petrol injection pump.

8.1.1 Fuel tank

The fuel tank is located remote from the engine. In a front-engined car, the fuel tank may be secured directly to the underside of the rear luggage compartment floor, generally by means of setscrews through a flanged connection. More recently, fuel tank design and installation have received attention to reduce the hazard which may follow a rear-end collision and spilling of fuel from a damaged tank. For this reason the fuel tank may now be located above the rear axle, so that it is better protected by the rear wheels and the surrounding body structure.

For rear-engined cars the fuel tank is usually either flange-mounted on to the front luggage compartment floor, or strap-mounted to the same compartment bulkhead.

Fuel tanks are traditionally fabricated from sheet metal, this being either corrosion-protected steel or aluminium alloy. The advantages of using the latter material include a saving in weight and inherently good resistance to corrosion, but any repairs are more difficult to effect. A fairly recent German development has been the moulded plastics fuel tank, which not only weighs less than a corresponding sheet metal tank, but possesses freedom from corrosion and also complies with modern safety requirements.

Most fuel tanks are fitted internally with a number of baffles which serve to minimise any violent surging of the fuel content during acceleration, braking and cornering. Otherwise, the fuel supply to the carburettor could be interrupted as a result of the fuel surging away from either the outlet pipe to the pump or, where an electrical pump is mounted inside the tank, the pump inlet port. The pivoted float of the fuel level indicator is

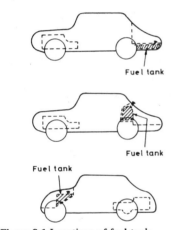

Figure 8.1 Locations of fuel tank

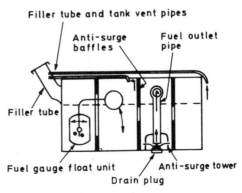

Figure 8.2 Features of fuel tank construction

generally shielded within a compartment formed by the baffle system. To prevent any accumulated sediment from entering the fuel supply system proper, the depth to which the outlet pipe enters the tank is always such that a low level of fuel remains at the bottom. For periodic removal of the sediment, a screwed drain plug is generally provided in the base of the tank.

A fuel filler tube with cap is connected towards the top of the tank and may simply be vented to the atmosphere, so that air can enter and take the place of fuel pumped from the tank. In the absence of suitable venting, a depression would be created in the tank that could not only interfere with normal fuel delivery, but also tend to collapse the tank. Another vent pipe may also be connected from the top of the fuel tank to a point near the top of the filler tube, so that air can leave the tank as fuel is pumped into it and thereby prevent an air lock.

In the early nineteen-seventies it was recognised in America that an important source of air pollution caused by the motor car was the evaporation of petrol from its fuel tank, especially on hot days. As a result, all cars sold there must now have their fuel tanks vented not to the atmosphere, but to a 'vapour recovery' system. When the engine is restarted following a period of shut-down, a mixture of filtered air and the previously stored vapours is then metered into the intake system for burning in the cylinders.

8.1.2 Fuel pump   A fuel pump is required to transfer petrol from the fuel tank to the carburettor since the installation level of the former is invariably lower than that of the latter and therefore precludes

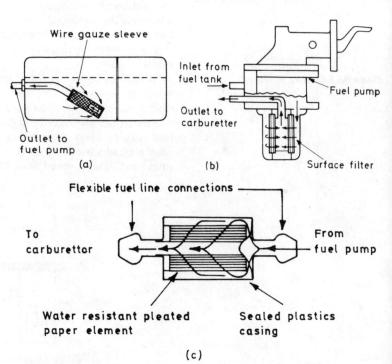

Figure 8.3 Examples of fuel filters: (a) coarse filter at tank outlet, (b) fine filter at pump outlet, (c) in-line fuel filter

the use of a 'gravity-feed' system as was used on early motor vehicles. In most carburettor fuel supply systems, a positive-displacement pump of the diaphragm type is operated either mechanically or electro-magnetically, the latter usually being described simply as 'electrical' pumps. An important requirement for both types of pump is that they must be self-priming. At this stage, we are to consider only the mechanical fuel pump as described in §§8.2 to 8.4.

8.1.3 Fuel filters   Filtration in the fuel delivery system to the carburettor serves two purposes, as follows:

(1) It must prevent foreign particles from becoming lodged in and interfering with the normal action of the fuel pump valve mechanism, and
(2) Similarly, it must protect not only the float-valve mechanism of the carburettor, but also its fuel-metering devices and internal passages.

In modern practice, fuel filtration is often performed by using a combination of coarse filters on the inlet side of the pump, and a fine filter on its outlet side. Fine filtration is not usually employed on the inlet side, because it may result in an increased tendency towards 'vapour locking' in the fuel line. Here it should be explained that vapour locking is most likely to occur at high operating temperatures, when any overheating of the fuel creates an excess of vapour in the supply system. This causes either partial or complete starvation of the fuel supply to the carburettor with accompanying misfiring and stalling of the engine.

A relatively coarse surface filter or strainer, in the form of either a circular screen or an extended sleeve of wire gauze, is sometimes located in the fuel tank itself at the entrance to the fuel line inlet. All fuel leaving the tank and entering the delivery system must therefore first pass through this filter. The fuel pump typically incorporates a built-in surface filter at the inlet to its pumping chamber. In mechanically-operated pumps, the filter often takes the form of a circular wire gauze screen.

In past practice, it was customary to afford additional protection for the carburettor by merely providing a relatively less coarse strainer at its fuel inlet, such as either a sleeve or a thimble gauze screen. More recently, the trend has been towards finer filtration for the fuel entering the carburettor, with the object of minimising flooding complaints. A fine depth-type filter of specially processed, water-resistant, pleated paper may therefore be incorporated in the pipeline between the pump and the carburettor, or before the outlet of the pump itself. The in-line type of filter is usually replaced in its entirety at appropriate servicing intervals.

8.1.4 Carburettor float chamber   The purpose of the carburettor 'float chamber' is to act as a constant-level reservoir for the fuel required by the 'mixing chamber'. To regulate the amount of fuel admitted to the float chamber a 'float valve' mechanism is used. This consists of either a brass, a rubber or a plastics float that actuates a 'needle-valve', the seating for which is contained in the chamber

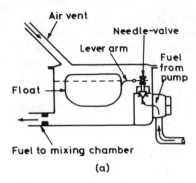

(a)

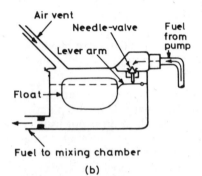

(b)

Figure 8.4 Carburettor float chambers: (a) with bottom fuel feed, (b) with top fuel feed

body or its cover. For accurate control of the fuel level, the needle-valve has a tapered tip to allow progressive opening and closing of the fuel inlet passage. In some cases a light compression spring is interposed between the needle valve and its end that contacts the float arm, so as to reduce vibration effects. A rubber-tipped needle-valve may sometimes be used instead of the all-metal type, the resilient tip permitting any fine particles to become temporarily embedded. Hence by ensuring that the needle-valve does not leak when closed, the carburettor is less likely to suffer from flooding.

The needle-valve is sited either co-axially above the float with which it is in direct contact or, more usually, alongside the float from which it is indirectly actuated by a lever arm. An advantage of the latter arrangement is that it provides a leverage in favour of the float, so as to enable a larger needle-valve area to be used where a greater through-flow of fuel is demanded. The lever arm is pivoted between the float and the needle-valve to suit a bottom fuel feed to the chamber, whereas it is pivoted at the opposite end to the float for a top fuel feed.

In operation, the float descends to open the needle-valve when the fuel level in the chamber falls. Conversely, as additional fuel flows into it, the float ascends by virtue of its buoyancy. When the desired level is reached, the needle-valve closes firmly on to its seating so that no further fuel can enter the chamber. To meet a continuous demand for fuel, the float mechanism remains open by an amount just sufficient to maintain the fuel level practically constant. The float chamber is ventilated either externally to the atmosphere or, as is now generally preferred, internally to the air intake of the carburettor.

8.1.5 Fuel pipelines

An obvious requirement of the fuel pipelines connecting the tank, pump, filter and carburettor is that they must be pressure-tight. If this condition is not maintained any leakage of petrol on the delivery side of the pump presents a fire hazard, while the entry of air on the intake side will reduce pumping capacity. Rigid metal pipelines are used for those parts of the system attached either to the body structure or to a separate chassis frame. Their installation is generally such that they are positioned away from the heat of the exhaust system and also afforded some protection from stones flung up beneath the car from the wheels. Usually the fuel pipelines are supported in small metal clips that may be lined with rubber, which not only protect the pipes against vibration, but also isolate the car interior from transmitted pump noise. A flexible hose is used for that part of the fuel pipeline attached between the car structure and the engine, because of the latter being resiliently mounted.

8.2 MECHANICAL FUEL PUMP

This type of fuel pump basically comprises a pumping chamber, which contains a pair of spring-loaded inlet and outlet valves and is sealed by a spring-returned diaphragm constructed from rubberised-fabric layers or laminae. Possessing very low inertia, the valves are of simple 'plate' type that makes them especially suitable for sealing against the low delivery pressures involved.

The oscillating or pumping motion of the diaphragm is derived from an eccentric cam, which is usually embodied on the engine camshaft. This eccentric motion of the cam is converted into one of oscillation at the diaphragm, through the medium of a pull-rod, since the latter is working in opposition to the diaphragm compression return spring. At its lower end the pull-rod is attached to a rocker-arm follower, which is always in contact with the eccentric cam. This permanent

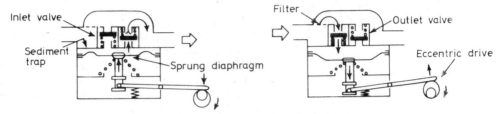

Figure 8.5 Operating principle of mechanical fuel pump

contact is obtained by separately spring-loading the rocker arm on to the cam and makes for quiet operation of the pump. In earlier practice, the rocker-arm usually took the form of an articulated lever assembly, but in modern examples the rocker arm is often made rigid with a forked end that simply intercepts a clearance collar at the lower end of the pull-rod. The reason for the apparent complication of the connection between the rocker-arm and pull-rod will be made clear in §8.4.

In operation, the intake stroke is performed by the diaphragm being retracted from the pumping chamber by the motion of the rocker-arm and pull-rod. The inlet valve thus opens as a result of the depression created in the pumping chamber. Since the fuel tank is vented either to the atmosphere or to a vapour recovery system, the resulting pressure difference in the fuel

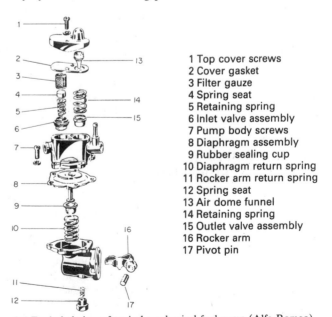

1 Top cover screws
2 Cover gasket
3 Filter gauze
4 Spring seat
5 Retaining spring
6 Inlet valve assembly
7 Pump body screws
8 Diaphragm assembly
9 Rubber sealing cup
10 Diaphragm return spring
11 Rocker arm return spring
12 Spring seat
13 Air dome funnel
14 Retaining spring
15 Outlet valve assembly
16 Rocker arm
17 Pivot pin

Figure 8.6 Exploded view of typical mechanical fuel pump (Alfa Romeo)

pipeline causes fuel to flow into the pump from the tank. At the end of the intake stroke the inlet valve closes, following which the outlet valve opens as the diaphragm advances into the pumping chamber on the delivery stroke. This is accomplished solely under the influence of the diaphragm return spring. Fuel is then being supplied under pressure to the carburettor float chamber.

8.3 FUEL PUMP DELIVERY PRESSURE

From the description just given of fuel pump operation, it will be evident that the main factor controlling delivery pressure is the strength of the diaphragm return spring, which is therefore chosen to suit the carburation requirements of a particular engine. In this context it is important that the diaphragm is not subjected to any stretching on its downward stroke, since this action would add to the force exerted by the diaphragm spring. A too high fuel pump pressure can result in the needle-valve being held off its seat in the carburettor float chamber, which in turn raises the fuel level and increases fuel consumption.

In practice, the fuel pump delivery pressure normally lies in the range 25–$50 \, kN/m^2$. It is generally highest at engine idling speed, when flow is least, and lowest at maximum speed, when flow is greatest.

8.4 FUEL PUMP DELIVERY REQUIREMENTS

Although the pump must be capable of delivering sufficient fuel under all conditions of engine operation—indeed, it must be more than capable so that it can provide quick priming and also handle any vapour present in the fuel pipeline—the quantity of fuel delivered must be only as much as the carburettor needs. This means that the pump must cease to deliver fuel when the needle-valve closes in the carburettor float chamber. At this point, the existing pressure in the fuel pipeline between the pump and the carburettor can produce an opposing force sufficiently high to balance the spring load on the diaphragm.

It is for this reason that the rocker-arm must embody some form of 'free-wheeling' facility, so that its normal motion can continue without disturbing the retracted diaphragm. This is achieved by ensuring that only on the intake stroke of the diaphragm is there a positive connection between the pull-rod and the rocker-arm.

8.5 THE SIMPLE FIXED-CHOKE CARBURETTOR

The process of delivering a combustible charge to the cylinders of a petrol engine is termed 'carburation'. Strictly speaking, it begins at the air and the fuel intakes and ends at the exhaust outlet, but it is customary to consider the fuel and the exhaust systems as separate entities. The purpose of the carburation system is therefore to meter fuel into the incoming air stream, in accordance with engine speed and load, and to distribute the mixture uniformly to the individual cylinders. Carburation is performed in most engines by means of one or more carburettors, although a fuel injection system is preferred by some manufacturers and will be examined at a later stage.

For theoretically complete combustion, the engine must be

supplied with a homogeneous mixture of air and petrol in the chemically correct proportions of approximately 14 parts to 1 part by weight. This would normally be expressed as an 'air–fuel ratio of 14:1' and is what the fuel chemist would describe as the 'stoichiometric' (chemically correct) air–fuel ratio. In actual practice, it becomes necessary to deviate from this ideal mixture to compensate for certain shortcomings inherent in petrol engine operation. Suitable provision must therefore be made for an

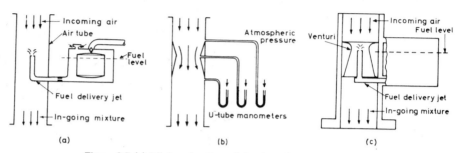

Figure 8.7 (a) Mixing chamber of simple carburettor, (b) pressure variations in venturi tube, (c) mixing chamber with venturi

excess of either air or petrol in the mixture, according to engine requirements, and this has led to the relatively complicated modern carburettor. At this stage, however, we shall be confining our attention to the main metering circuit of the fixed-choke carburettor.

8.5.1 The mixing chamber of the fixed-choke carburettor

The purpose of the mixing chamber is to 'measure' the incoming air flow and to meter into it the appropriate amounts of fuel. In its simplest form, the mixing chamber could consist merely of a plain air or 'choke' tube into which is placed a fuel delivery jet, so that the incoming air to the cylinders is induced to flow past it whenever the engine is running. Since a moving stream of air exerts less sideways pressure than when it is stationary, the fuel entering the delivery jet from the floatchamber is forced out into the air stream. This is because the pressure acting on the fuel supply in the float chamber remains at atmospheric and is therefore greater than that surrounding the jet. It is misleading to think in terms of the fuel being 'sucked' from the jet, rather it is 'pushed' from the jet due to a difference in pressures. However, the pressure difference so created would be hardly sufficient to overcome the resistance of the fuel in the jet to flow. In order to increase the depression over the jet, the air tube must be provided with a constricting passage known as a 'venturi'. This depends for its action on the 'Bernoulli' effect, so-named after the eighteenth century Swiss scientist Daniel Bernoulli, who wrote extensively on the subject of hydraulics.

Stated in simple terms, Bernoulli's principle is that the relationship between pressure and velocity in a flowing 'fluid' (in our case air) is such that the total energy possessed by the air flow is the same at every point along its path. The pressure of the air will therefore be reduced where its velocity is increased, since the gain in kinetic energy must be offset by a loss in potential energy. Because the same volume of air must flow through all portions of the carburettor air tube, it will be evident that the effect of introducing a venturi is to increase locally the air speed and cause a corresponding reduction in pressure over the fuel delivery jet.

Verification of this Bernoulli effect could be made by connecting a series of U-tube manometers into the venturi. Pressure gauges of this type work on the principle that the pressure in a column of liquid is directly proportional to its height. Hence, the difference in level of the liquid in the limbs of each U-tube is a measure of how much less the air pressure in the venturi is than atmospheric pressure. It is the pressure drop between the air intake and the venturi which, in effect, enables the carburettor to 'measure' the incoming air flow for the purpose of metering the fuel flow. Furthermore, the pressure drop serves to ensure that the jet delivers a finely-atomised spray of fuel into the air stream. To promote the best conditions for air flow, the venturi is provided with a smooth, rounded entrance and a gently tapering exit.

The mixing chamber also incorporates a 'throttle valve', which is positioned on the engine side of the venturi and is linked to the accelerator pedal by either a cable in a conduit or, now less commonly, a rods-and-levers system. This valve serves to regulate the amount of air and fuel mixture admitted to the engine cylinders and hence the power they develop.

8.6 LIMITATIONS OF THE SIMPLE MIXING CHAMBER

A simple mixing chamber of the form just described does, in practice, need considerable refinement if it is to satisfy the mixture requirements of the variable-speed engine. With increasing air flow through the venturi, a plain metering jet

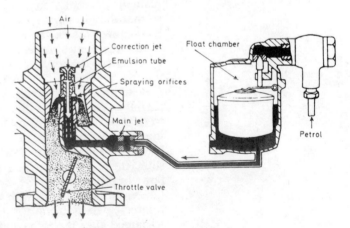

Figure 8.8 An air-bleed system of mixture correction

would tend to supply disproportionately greater quantities of fuel, resulting in an increasingly richer mixture. This is because the density of the air will vary accordingly to the pressure, whereas the fuel is virtually incompressible and does not share this characteristic. While this is not disadvantageous in supplying the rich mixture required for full-throttle performance, it is opposed to the requirement for a relatively weak mixture suitable for economical part-throttle cruising. To meet the latter requirement demands some method of correcting mixture strength.

8.7 MIXTURE STRENGTH CORRECTION

There are three basic methods that may be used to corrrect mixture strength in the fixed-choke carburettor. At this stage, we are required to examine only one of these which utilises a restricted air bleed and correction jet. This method is often quite simply referred to as the 'air bleed' system.

8.7.1 Restricted air bleed and correction jet

With this system of air bleed mixture correction, which continues to be used in many modern carburettors, a single fuel-metering orifice is employed to serve as the main jet. It is generally sited near to the base of a 'discharge nozzle' and is thus submerged in the fuel supplied to the float chamber. A perforated 'emulsion tube' is either incorporated within, or located remote from, the discharge nozzle and terminates in a restricted 'air bleed' or 'correction jet'. This is exposed to the incoming air flow, so that the fuel flow metered by the main jet is subject to a lesser depression than that existing at the venturi.

In operation, an increasing depression in the venturi promotes a greater flow of fuel from the discharge nozzle, the level in which falls accordingly. As a result, the fuel stream becomes emulsified by air admitted through the emulsion tube perforations via the restricted air bleed of the correction jet. Since more of the holes distributed over the length of the emulsion tube will be uncovered as the pressure drop across the venturi increases, a variable air bleed to the main metering circuit is thus effected. In other words, the mixture-enriching effect of the main fuel jet is balanced by the weakening effect of the restricted air bleed, so that a substantially uniform mixture strength is obtained.

8.8 THE PURPOSE OF THE AIR CLEANER AND SILENCER

Efficient removal of dust particles from the considerable quantities of air flowing into the engine is of vital importance. Otherwise, their presence would contribute to internal abrasive wear, especially between the working surfaces of the cylinders, pistons and rings. It has therefore been long established practice to install an air filter unit on the intake system of motor vehicle engines.

The air cleaner is also required to act as a silencer for the carburation system; that is, it must suppress engine induction noise to an acceptable low level. With small throttle openings, induction noise is generally of a high-frequency character and may be heard as a high-pitched 'hiss' from the carburettor intake. At large throttle openings, the major source of induction noise occurs as low-frequency or 'boom' periods, which arise from the implosion of the air and fuel charge into the cylinders.

Another function performed by all air cleaners is to act as a flame arrester, in the event of the engine backfiring through the induction system.

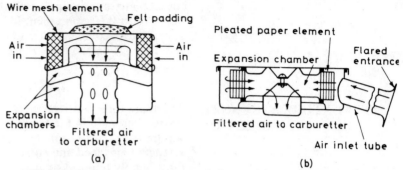

Figure 8.9 Two types of air cleaner and silencer: (a) early oil-wetted wire mesh type, (b) later dry paper element type

8.9 TYPES OF AIR CLEANER

Since a detailed knowledge of the cleaning and silencing functions of the different types of air cleaner is not required until a later stage, suffice it to mention here that dust particles may be removed from the incoming air stream either by their adhering to an oil-wetted surface, or being excluded by a depth filter of the paper element variety. The latter method represents established modern practice. For particularly arduous duty a centrifugal pre-filter may be used in conjunction with a depth filter.

9 Diesel engine fuel system

9.1 LAYOUT AND PURPOSE OF COMPONENTS

The motor vehicle diesel engine fuel system comprises the following services:

(1) Fuel tank
(2) Preliminary filter
(3) Fuel lift pump
(4) Main filter(s)
(5) Fuel-injection pump
(6) Injection-pump governor
(7) Fuel injectors
(8) Fuel pipe lines.

(1) *Fuel tank:* This is supported from the side of the chassis frame, as described in §1.1.8 of *Vehicle Technology 1*. It is located in a position as sheltered as possible from the cooling air flow past the vehicle, so as to afford some protection against wax crystals separating from the fuel oil and blocking the fuel system during severe winter operation.

   The fuel tank is usually provided with a coarse strainer gauze at its filler neck, a drain plug in its base, and a fuel gauge with either a remote cab reading or a direct tank reading, to indicate tank content. Connections are provided at the top of the tank for the fuel delivery pick-up and return leak-off pipes. Internal baffles are fitted to minimise surging of the fuel. The capacity of the diesel fuel-oil tank is usually several times greater than that of the car petrol tank. For medium to heavy commercial vehicles it is typically in the region of 300 litres.

(2) *Preliminary filter:* This is situated in the fuel pipe line between the tank and the fuel lift pump. Current thinking favours only a simple 'sedimenter' filter, since the presence of a fine wire gauze at this point could aggravate any

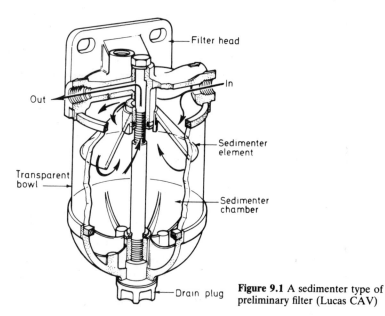

Figure 9.1 A sedimenter type of preliminary filter (Lucas CAV)

blocking tendency caused by fuel-oil waxing. The purpose of a sedimenter filter is to remove the larger droplets of water and abrasive matter that may be present in the fuel oil being pumped from the tank. Water in the fuel oil is to be avoided because it can interfere with the proper lubrication of the fuel-injection equipment. The undesirable effect of abrasive matter in the fuel oil is dealt with later.

(3) *Fuel lift pump:* Since the fuel tank is mounted below the level of the engine fuel-injection pump, a fuel lift (or transfer) pump is required to maintain a small constant head of fuel oil in the feed gallery for the injection pumping elements. Although various types of positive-displacement lift pumps may be encountered in diesel engine practice, those used for automotive applications are typically of either the 'plunger' or 'diaphragm' variety. Both are provided with hand-priming levers so that fuel oil can be forced into the system to vent it of air, without turning the engine.

(4) *Main filter(s):* Contamination of the fuel oil by abrasive matter is the great enemy of the diesel fuel injection system, and, for that matter, any other piece of hydraulic equipment. The effect of abrasive matter in poorly-filtered fuel oil reaching the injection pump and injectors is to damage their highly finished and selectively assembled mating parts. This can result in reduced performance, poor starting and irregular idling of the engine, because of decreased delivery from the injection pump. In the case of the injectors, their faulty spraying and leakage can increase both fuel oil consumption and exhaust smoke. It is therefore

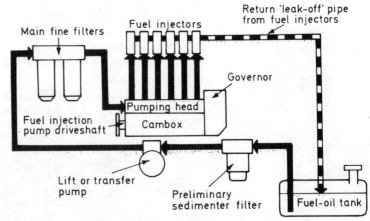

Figure 9.2 Simple representation of diesel engine fuel injection system (in-line pump)

necessary to incorporate one or more fine filters between the lift and injection pumps. The main filter(s) is deliberately sited so that it can receive heat from the engine, again to minimise fuel-oil waxing problems.

(5) *Fuel injection pump:* Its purpose is to deliver fuel oil at higher than combustion chamber pressures, in minute quantities exactly related to the amount of power required. Furthermore, the delivery of the fuel oil must be timed to

occur at exactly the required moment in the engine operating cycle. At this stage our attention will be confined to the multi-element 'jerk' pump, in which the pumping elements and their delivery valves are arranged inline and operated by a camshaft within the pump body. This type of fuel-injection pump is either flange- or platform-mounted alongside the cylinder block of in-line engines and between the cylinder banks of V engines. It is gear-driven from the engine camshaft through a 1:1 ratio, since, like a petrol engine ignition distributor, the fuel-injection pump in the four-stroke cycle must also be driven at half engine speed.

(6) *Injection pump governor:* In the conventional petrol engine, the carburettor meters fuel into the incoming air stream, which is throttled in accordance with power requirements. The marriage of fuel and air thus takes place at a single source and the process may be regarded as self-regulating. This characteristic is not shared by the diesel engine, since its fuel-injection system is entirely divorced from the incoming air to the cylinders and power is controlled solely by varying the quantity of fuel injected into the cylinders. For diesel engines, therefore, it becomes necessary to superimpose a 'governor' control to provide automatic speed regulation, relative to any set position of the accelerator pedal. The governor mechanism is combined with the fuel-injection pump and will be described at a later stage.

(7) *Fuel injectors:* These represent the final link in the fuel-injection system and control the passage of metered quantities of fuel oil to the engine combustion chambers. They are required not only to deliver a finely divided spray of fuel, but also to distribute it uniformly through the very hot, compressed and turbulent air in the combustion chambers. Each cylinder has a separate fuel injector which, depending upon application, may be flange-mounted, clamped or screwed into the cylinder head.

(8) *Fuel pipe lines:* The fairly generous bore size of the low-pressure tubing used on the supply side of the injection pump is intended to minimise both pressure drop and the energy required to operate the lift pump. On the delivery

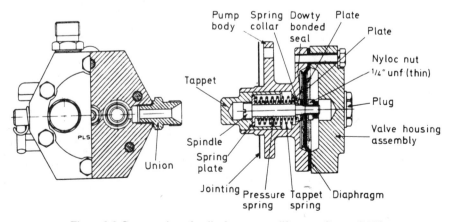

Figure 9.3 Cross-section of a diaphragm-type lift pump (Lucas CAV)

side of the injection pump the exceedingly small bore size of the high-pressure tubing is related to the diameter of the pumping elements, and its greater wall thickness to the high injection pressure being handled. It is also important that the delivery tubing should be of minimum, but nevertheless equal, length from the pump to the individual injectors. Seamless steel tubing is used for the fuel pipe lines, which should have only large-radius bends and also be clamped to avoid fracture from vibration. The end connections may be effected through either conical sleeves or banjo unions.

9.2 TYPES OF LIFT PUMP

The 'single-diaphragm' type of lift pump is very similar to that found in petrol engine fuel systems. It may be mounted on the engine crankcase or on the fuel-injection pump and takes its drive from the camshaft of either. For greater capacity a 'twin-diaphragm' arrangement can be used, which provides a double-acting pump and embodies two sets of inlet and outlet disc valves in a central partition. This type of diaphragm pump is driven directly from the camshaft of the fuel-injection pump on which it is mounted.

The 'plunger' type of lift pump is comprised basically of a spring-energised piston that reciprocates in a pumping chamber containing the inlet and outlet valves. A diaphragm is arranged in series with the pumping piston, its purpose being to seal against any fuel oil that leaks past the piston. In a more recent version with improved sealing, a second diaphragm has been introduced for retaining oil in the fuel-injection pump cambox. The pumping piston is operated either through a bell-crank lever or, as in current practice, a direct-acting tappet with roller follower, which bear against an eccentric on the camshaft of the fuel-injection pump. This type of lift pump generates a higher pressure. A lost-motion facility is incorporated in the operating mechanism of these different types of lift pump so that they can 'free-wheel' when the force on the diaphragm or the plunger due to back pressure exceeds that of the energising spring. In earlier practice this feature was not present and a pressure relief valve had to be included in the system, together with additional piping, so that surplus fuel could be returned to either the tank or the inlet side of the lift pump.

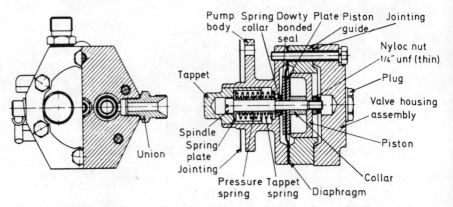

Figure 9.4 Cross-section of a plunger-type lift pump (Lucas CAV)

9.3 LIFT PUMP DELIVERY PRESSURES

As in the case of the petrol engine fuel system pump, the controlling factor for delivery pressure in the diesel lift pump is the strength of the energising spring acting on either its diaphragm or plunger. The former operates at a lower pressure than the latter with maximum values being in the regions of 30 and $100 \, kN/m^2$ respectively. There is, of course, a reduction in pump output as the back pressure on the pump increases.

9.4 TYPES OF FUEL FILTER

The secondary fuel-oil filters used between the lift pump and the injection pump may be classified according to the direction of flow through their filtering media, as follows:

(1) Down-flow
(2) Up-flow
(3) Cross-flow.

(1) *Down-flow:* Filters of this type are arranged to provide what is termed an 'agglomerator' flow. This is based on the principle that when fuel oil containing any fine water droplets is passed through a porous filter element, they will join together, or 'agglomerate', into larger drops which can then be collected by sedimentation. In operation the incoming unfiltered fuel passes downwards through the filter element, which comprises a V-form paper roll construction, and into the base of the filter body. It is then displaced upwards through the annular space created by the central tube of the filter element and the filter body securing bolt. Finally, the filtered fuel leaves at the outlet connection of the filter head. In this manner, the abrasive particles are retained by the porous filter element and the water content is collected in the base of the filter.

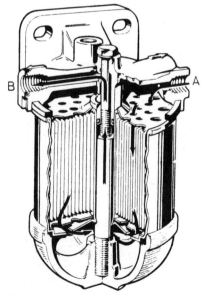

Figure 9.5 A down-flow type of fuel-oil filter (Lucas CAV). A: unfiltered fuel oil inlet. B: filtered fuel oil outlet

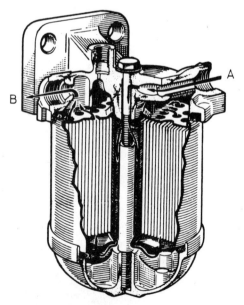

Figure 9.6 An up-flow type of fuel-oil filter (Lucas CAV). A: unfiltered fuel oil inlet. B: filtered fuel oil outlet

(2) *Up-flow:* A purely 'filter' rather than 'agglomerator' flow is provided with this filtering arrangement. Its action is such that the incoming unfiltered fuel first passes downwards through the annular space created as before by the central tube of the filter element and the filter body securing bolt. The fuel is then displaced upwards through the filter element to leave immediately from the outlet connection of the filter head.

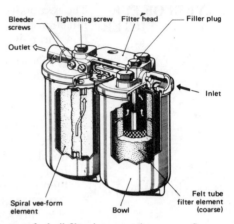

Figure 9.7 Two-stage fuel-oil filter incorporating a cross-flow coarse element

(3) *Cross-flow:* This corresponds to the more familiar filtering action described earlier in connection with the engine lubrication system; that is, the incoming unfiltered fuel passes inwards through the filter element, which in this case assumes a radially-pleated form. The filtered fuel from the interior space of the element then leaves via the outlet connection of the filter head.

9.5 THE IN-LINE FUEL INJECTION PUMP

Early attempts by Rudolph Diesel to use mechanical methods of injecting fuel oil into the combustion chamber proved unsatisfactory, no doubt because of the relatively crude injection equipment then available to perform such a precision task. He therefore resorted to what was known as 'air-blast' injection. In this system a compressor and storage tank supplied air at a pressure much higher than that existing in the engine cylinder, as a means of injecting and atomising the fuel oil. This system proved quite satisfactory and remained accepted practice for large marine and stationary diesel engines until the early nineteen-thirties.

It was also about this period that the diesel engine was being developed for higher speeds and lighter weight to make it suitable for road transport applications. The air-blast injection system had therefore to be discarded on the grounds that not only was it too heavy, bulky and costly, but also because the air compressor robbed the engine of up to 10% of its power output. A return was therefore made to mechanical methods of injection, albeit of more sophisticated form, in which extremely small amounts of fuel oil (in the order of one-tenth of a cubic centimetre) are forced through spray nozzles under high hydraulic pressures from a 'fuel injection pump'. The first commercially successful pump of this type was the 'port-controlled' jerk pump produced by Robert Bosch of Germany, which was later adopted with detail variations in design by other manufacturers in both Britain and America.

9.5.1 Construction and operation   The in-line fuel injection pump comprises the following parts:

(1) *Pump housing:* This includes the 'cambox' and the 'pumping head', the former to support the camshaft, tappet assemblies, control rod and governor mechanism, and the

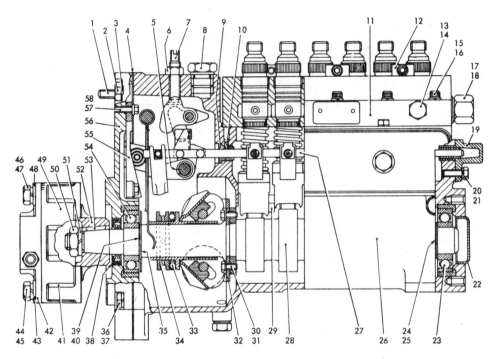

1 Stud
2 Nut
3 Spring washer
4 Gasket
5 Woodruff key
6 Nut
7 Maximum fuel stop screw
8 Oil filler plug
9 Control rod bush
10 Groverlok pin
11 Pump body
12 Clamp
13 Air vent screw
14 Joint washer
15 Screw
16 Spring washer
17 Fuel inlet adaptor
18 Joint washer
19 Control rod cover
20 Screw

21 Spring washer
22 Expansion plug
23 Ball bearing
24 Shim (0.1 mm)
25 Shim (0.2 mm)
26 Pump unit housing
27 Control rod
28 Camshaft
29 Tappet locating T-piece
30 Backplate screw
31 Tab washer
32 Backplate
33 Thrust bearing
34 Thrust pad
35 Baffle washer
36 Screw
37 Spring washer
38 Oil seal
39 Shim (0.1 mm)
40 Shim (0.2 mm)

41 Insert
42 Driving flange
43 Clamp plate
44 Dowel screw
45 Spring washer
46 Screw
47 Spring washer
48 Clamp plate
49 Dog flange
50 Camshaft nut
51 Spring washer
52 Pump flange
53 Woodruff key
54 Ball bearing
55 Ramp
56 Governor cover
57 Screw
58 Lock washer

Figure 9.8 A modern fuel injection pump for a six-cylinder diesel engine (Lucas CAV)

latter to contain the pumping elements and delivery valves. It may be produced either as a one-piece housing in aluminium alloy, or a steel pumping head may be separately bolted to an aluminium alloy cambox. Whichever form of construction is used, the overriding consideration is for maximum rigidity of the pump housing, since this can affect both pumping efficiency and operating noise.

(2) *Governor housing:* This usually takes the form of an enlarged extension of the cambox and is closed by a suitable end cover.

(3) *Camshaft and tappet assemblies:* These operate in conjunction with the plunger return springs to provide the reciprocating motion for the pumping elements. The camshaft is supported by heavy-duty rolling bearings, ball or opposed tapered rollers, and is of rugged proportions to withstand without deflection the heavy loads imposed upon it. Either or both ends may have standard tapers with keyways which provide connections for a driving coupling or gear and the governor. Each spring-returned pump plunger is lifted by its cam through a tappet assembly, incorporating a roller, bush and pin to ensure smooth rolling contact with the cam. A tappet adjustment for 'phasing' the pump is provided and is an operation that will be described at a later stage.

(4) *Pumping elements:* Each of these comprises a barrel and plunger assembly, both components being made from steel. In providing a sliding fit, the barrel and plunger are matched to such fine limits as to eliminate the need for any other means of sealing between them. Here it should be appreciated that we are thinking in terms of a working 'clearance' of about 1.5 μm! The barrel is located in the pumping head either by a locking screw or serrations and retained endwise by the delivery valve holder. Sandwiched between these parts is the delivery valve guide. Fuel oil 'inlet' and 'spill' ports are provided in the upper portion of each barrel and register with corresponding galleries in the pumping head.

Each plunger is shaped with a sloping edge or helix at its head portion, so that although the plunger operates over a

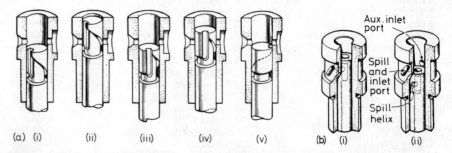

Figure 9.9 Showing the action of two types of pumping element. (a) Helix and slot: (i) commencement of delivery stroke, (ii) termination of fuel injection under full load, (iii–iv) commencement of delivery stroke and termination of injection under partial load, (v) position of plunger for stopping engine. (b) Helix and drillings: (i) plunger at commencement of spill, (ii) plunger at bottom of stroke

constant stroke, the quantity of fuel delivered can be varied by partial rotation of the plunger. That is, it depends on the radial position of the plunger helix relative to the spill port in the barrel. Hence, the earlier the helix registers with the spill port, as the cam lifts the plunger, the lesser will be the quantity of fuel delivered. No fuel is delivered at all, of course, if the helix uncovers the spill port over the entire stroke of the plunger. Since it is necessary to connect the space above the top of the plunger to the spill port, the plunger head must also be provided with either a vertical slot in its sidewall, or a central hole with a radial duct. An advantage of the latter method is that there is less reduction in contact area, so that wear and leakage are reduced. The actual method of fuel metering by 'port-control' was not invented by Robert Bosch, but was conceived earlier by another German, Carl Pieper, in 1892.

9.6 PUMPING ELEMENT CONTROL

There are two basic methods used for simultaneously turning the reciprocating plungers to control fuel delivery, these being known as 'rack-and-pinion' and 'fork-and-lever'.

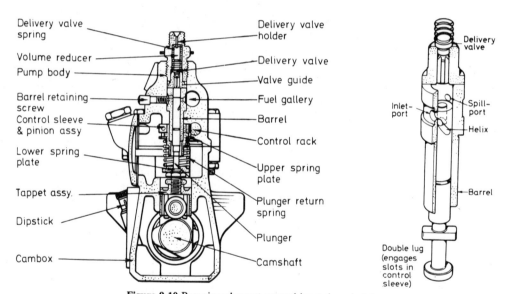

Figure 9.10 Pumping element control by rack-and-pinion

9.6.1 Rack-and-pinion

This was the earliest, and for many years, the most widely used method of pumping element control. In this construction each plunger can be partially rotated through the medium of lugs formed near its lower end. These engage with slots in an extended sleeve that surrounds, and can turn on, the lower part of the barrel. Clamped to the upper part of the sleeve is a quadrant pinion that meshes with a toothed portion of the sliding control rod. It thus follows that any endwise movement of the latter will produce a partial rotation of the quadrant pinion, together with slotted sleeve and likewise the reciprocating plunger.

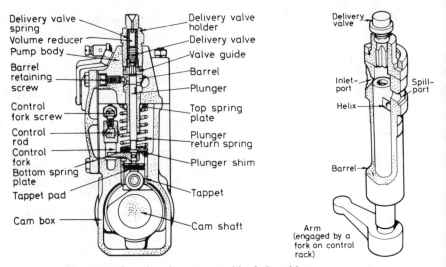

Figure 9.11 Pumping element control by fork-and-lever

9.6.2 Fork-and-lever

With this method of pumping element control the slotted sleeve, quadrant pinion and toothed control rod are dispensed with and replaced simply by a radial lever attached to either the foot or the waist of the plunger and engaging a vertical fork carried on a square section control rod. Therefore any sliding movement of the latter will cause the forks to swing the outer ends of the plunger levers and thus partially rotate the plungers. The depth of slot in the fork is, of course, sufficient to accommodate the reciprocating movement of the plunger and its lever.

The main advantage generally claimed for this particular method of pumping control is that the sources of backlash in the mechanism are reduced from two with the rack, quadrant and slotted-sleeve arrangement to one between the fork-and-plunger lever, and even this is acting at a large radius to reduce the effect of backlash. There is also less possibility of any misalignment friction occurring and causing sluggish control-rod movement.

9.7 PUMPING ELEMENT DELIVERY VALVE

A spring-loaded delivery valve is fitted at the head of each pumping element and serves a twofold purpose, as follows:

(1) To act as a 'non-return' valve that isolates the high-pressure from the low-pressure regions of the pumping element, thereby preventing fuel from re-entering the barrel when the plunger retreats on its intake stroke.
(2) To serve as an 'unloading' valve that creates additional space or 'dead volume' on the delivery side of the pumping element, thereby ensuring a sudden drop in line pressure and clean termination of spray from the injector.

Although the delivery valve may appear to be a fairly ordinary sort of non-return valve, it does in fact represent a very elegant solution to what was originally a difficult problem in hydraulics, and all credit for this must go to Axel Danielson of the Atlas Diesel Company in Sweden. In 1924 he introduced a 'retraction' type delivery valve basically resembling a taper-seating poppet valve, but which had its guide stem effectively divided into two parts of the same diameter. The shorter upper part was formed as a separate collar, while the lower part was fluted to allow the passage of fuel.

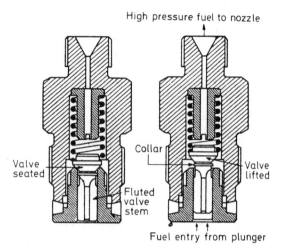

Figure 9.12 Atlas type of pumping element delivery valve

It acts in a manner similar to an ordinary non-return valve, but with the important difference that, when closing, the effect of the collar entering the valve seating bore is to produce an automatic increase in volume in the pipeline to the injector. Hence, the accompanying sudden drop in pressure ensures that the fuel spray from the injector is cleanly terminated when its needle valve closes. Otherwise, there is the likelihood of 'dribble' occurring from the fuel injectors, which was one of the problems encountered with early jerk pump fuel-injection systems.

9.8 FUEL INJECTORS

Since the pumping elements in combination with the delivery valves not only supply the necessary pressure for injecting fuel oil into the engine combustion chambers, but also control both the timing and period of injection, it might seem at first that a fuel injector need comprise nothing more than a simple spraying hole. This would be feasible were it not important that full pressure is required at the beginning of fuel injection and, as previously mentioned, clean termination of fuel spray at the end of injection. For these reasons the fuel injector must incorporate a spring-loaded valve.

The inwardly-opening 'differential needle valve' type of fuel injector has long been established practice and was originated by Thornycroft in this country as long ago as 1908. In present-day constructions, the larger-diameter guiding portion of the needle-valve is lapped into the nozzle body to render the bearing surfaces virtually self-sealing, while its smaller-diameter lower end has a conical seating to seal the nozzle orifice.

Operation of the fuel injector is such that the rising delivery

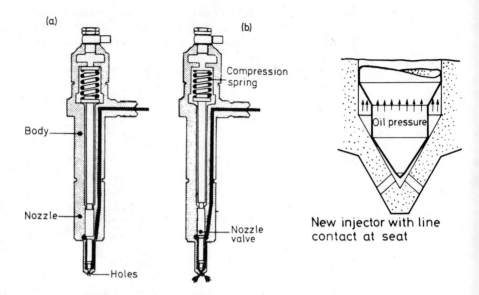

Figure 9.13 Construction and operation of a fuel injector (Lucas CAV). (a) Fuel spray terminated, (b) fuel spray initiated

pressure of the fuel acts initially upon the annular area above the valve tip, until the spring pre-load on the needle-valve is overcome. At this stage, the delivery pressure additionally acts upon the valve tip, so that the needle-valve rapidly and unhesitatingly attains its full lift and opening. Fuel is then sprayed from the nozzle until the pressure falls and the spring loading returns the needle valve smartly to its seating. This action is assisted by the delivery valve, as mentioned above.

Since there must inevitably be some leakage past the sliding needle-valve, and indeed this is necessary for lubrication purposes, a 'leak-off' pipe is provided to return surplus fuel to the supply tank.

There are various types of fuel-injector nozzle manufactured, the purpose of which will be better understood when diesel engine combustion chamber design is considered at a later stage. It should nevertheless be appreciated that the fuel injector leads a particularly arduous life. A leading manufacturer of diesel engine fuel injection equipment, Lucas CAV Limited, state in connection with fuel injectors that to perform their duty consistently and efficiently, nozzles are manufactured to tolerances as small as 1.5 μm in materials which can withstand high pulsing fuel pressure of up to 60 000 kN/m^2 and temperatures of over 400°C.

10 Engine cooling systems

10.1 AIR COOLED ENGINES

From the viewpoint of converting heat into mechanical energy, it follows that if each piston accomplished its power stroke starting at the temperature of combustion, such an engine would in theory be highly efficient. To achieve this in practice, however, would entail unacceptably high operating temperatures with adverse effects on both engine lubrication and materials.

Excessively high operating temperatures would cause breakdown of the lubricating oil films, resulting in undue wearing and possible seizure of the working parts. The behaviour of metals at high temperatures also differs from that at normal temperatures and can produce a condition known as 'creep', in which the metal deforms slowly and continuously at a constant stress.

For these reasons the engine must be provided with a system of cooling, so that it can be maintained at its most efficient practicable operating temperature. This generally means that the temperature of the cylinder walls should not exceed about 250°C, whereas the actual temperature of the cylinder gases during combustion may reach ten times this figure.

Conversely, there is no merit in operating the engine too cool since this would reduce thermal efficiency and therefore increase fuel consumption. It would also increase oil dilution and hasten corrosion wear of the engine.

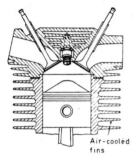

Figure 10.1 An air-cooled cylinder and head

10.1.1 Circulation of cooling air

It is a common experience that heat always flows from a hotter to a colder substance, and a physical principle that the greater the temperature difference between them the more rapidly is the heat transferred. With air cooling, the engine structure is directly cooled by inducing air to flow over its high temperature surfaces. These are finned to present a greater cooling surface area to the air, which in non-motorcycle applications is forced to circulate over them by means of a powerful fan. The car engine structure is almost entirely enclosed by sheet metal ducting, which incorporates a system of baffles.

These ensure that the through-flow of air is properly directed over the cooling surfaces of the cylinders and cylinder heads. To maintain uniform temperatures, the air is forced to circulate around the entire circumference of each cylinder and its cylinder head, the direction of flow being along the cooling fins. These are greatest in number, consistent with providing a sufficient area of flow, on the high temperature surfaces of the cylinder head in the region of the exhaust valve.

The complete system forms what is known as a 'plenum chamber' in which the internal air pressure is higher than that of the atmosphere. Finally, the heated air is discharged from the plenum chamber to the atmosphere, or re-directed to heat the car interior.

10.2 AIR-COOLING FAN ASSEMBLY

The forced circulation of air around the engine is generally provided by a centrifugal fan or impeller, which rotates in a spiral-shaped housing. This type of fan is capable of overcoming the appreciable resistance offered to the air as it flows around the ducted and finned cylinders and cylinder heads. The fan is driven by a V-belt and pulley system from the engine.

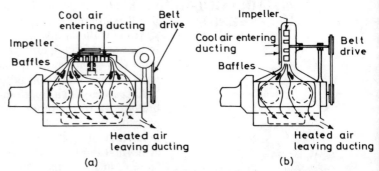

Figure 10.2 Air-cooling systems with (a) vertical bearing axis and (b) horizontal bearing axis for impeller

During operation the air enters at the 'eye' of the impeller via an inlet housing, and flows between each pair of blades. Centrifugal force thus acts upon the enclosed air, which is then discharged under pressure in a radial direction. To increase fan efficiency the impeller blades may be curved backwards instead of being straight. The fan housing is also fitted with an 'inlet ring' to minimise any recirculation of air at the point of entry.

10.2.1 Air-cooling throttle valve   This regulates the quantity of air entering the cooling fan, in accordance with engine cooling requirements. It usually takes the form of a movable 'throttle-ring', which acts as a baffle surface at the entrance to the inlet housing.

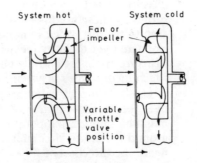

Figure 10.3 Air-cooling throttle valve

At normal running temperatures, the throttle ring is retracted from the eye of the impeller to allow free access for the incoming air. When the engine is cold, the throttle ring is advanced towards the eye of the impeller, thus reducing the quantity of air circulated around the engine.

10.2.2 Air-cooling thermostat   Automatic operation of the cooling fan throttle valve is effected by a thermostat, which therefore serves to regulate the rate of engine cooling. It is linked mechanically through a simple leverage system to the throttle valve, so that adequate movement is transmitted to the ring within the available

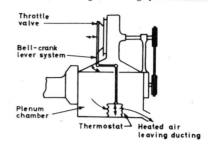

Figure 10.4 Air-cooling thermostat

working stroke of the unit. The thermostat is generally installed in the hot air duct leading from the lower part of the plenum chamber.

10.3 THE THERMO-SYPHON WATER-COOLING SYSTEM

Motor vehicle engines are designed either for 'direct' cooling by air, as just described, or more commonly for 'indirect' cooling by air through the medium of water. More effective cooling is generally attributed to the latter system, since heat is transferred more readily from the engine metal surfaces to water than it is to air. This is because the specific heat of water is greater than that of air. In other words, if equal masses of water and air each receive the same quantity of heat, then the water will experience the least rise in temperature. Heat transfer will therefore be more rapid as a result of the higher mean temperature difference between the metal surfaces and the water in contact with them.

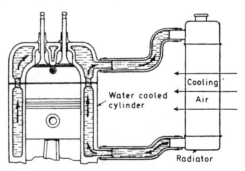

Figure 10.5 The thermo-syphon system of water-cooling an engine

Although it follows that more rapid warm-up of the engine can be attained with air cooling, by the same token the air-cooled engine will cool down more rapidly, which is less advantageous. Water cooling can, in fact, be said to possess greater 'thermal inertia', which helps to maintain a more nearly constant operating temperature and this is its chief advantage.

At this stage, it is only necessary to acquire a knowledge of the simplest form of engine water-cooling system, known as 'natural' or 'thermo-syphon' circulation.

10.3.1 Essential components of the thermo-syphon system

The simple thermo-syphon cooling system was used in many early motor vehicles up to the late nineteen-thirties and consisted of the following components:

(1) *Water jacket:* This was formed partly in the cylinder block and partly in the cylinder head and surrounded the cylinder walls, combustion chambers and valve ports. It was provided with a lower inlet and an upper outlet connection, the latter sometimes being called the 'header pipe'.

(2) *Radiator:* The purpose of the so-called radiator is to dissipate the engine heat rejected to the coolant. This it does mainly by conduction and convection, rather than by radiation. It originally consisted of an upper 'header tank' that received the heated coolant from the engine; a matrix or core that served to disperse the down flow of coolant into fine streams and also provided for a through-flow of cooling air; and a lower collector tank from which coolant was returned to the engine jacket.

(3) *Flexible connections:* Rubber hoses with clips were used to connect the cooling jacket header pipe to the inlet pipe of the radiator header tank and the outlet pipe of the radiator collector tank to the inlet elbow of the cooling jacket.

(4) *Other fittings:* The radiator header tank was fitted with a filler cap and an overflow pipe and the collector tank with a drain tap. Suitable mounting lugs were also provided at the top and bottom of the radiator.

(5) *Radiator fan:* In many, but not all, early thermo-syphon cooling systems a fan was mounted behind the radiator and driven by belt and pulleys from either the crankshaft or the camshaft. Its purpose was to assist the through flow of air normally resulting from the forward motion of the vehicle and also create an air flow when the vehicle was stationary with the engine idling.

10.3.2 Operating principle of the thermo-syphon system

Upon starting the engine from cold, the following sequence of events occurs:

(1) Cold water in the jacket begins to absorb surplus heat from the engine and expands slightly.

(2) Since the heated water becomes less dense than the cold water in the jacket, it rises to the header pipe and then flows upwards into the header tank of the radiator.

(3) This heated water then begins to be forced down in fine streams through the radiator matrix, as a result of more water being heated in the jacket and rising behind it.

(4) The heated water passing through the radiator is then cooled by the through-flow of cooling air, so that it becomes denser again and readily sinks to the collector tank of the radiator.

(5) Cooled water is then raised from the collector tank to the inlet of the jacket as a result of the pressure difference created by other cooled water sinking in the radiator and heated water rising in the jacket again.

The rate at which water circulates in the simple thermo-syphon cooling system is not in proportion to engine speed, but to the heat output or load placed on the engine. As the engine load increases the circulation becomes more rapid, mainly because the formation of small steam bubbles have the effect of further reducing the density of the coolant being heated.

Purely as a matter of historical interest, perhaps the best

remembered application of thermo-syphon cooling was to the legendary Model 'T' Ford car, which remained in production for nearly 20 years up to 1927. Of great simplicity in construction, this was the first car to be mass-produced in the true sense of the word and also the one about which Henry Ford made his famous remark that – any customer can have a car painted any colour that he wants so long as it is black.

10.4 THERMO-SYPHON *v* FORCED-CIRCULATION COOLING

In long-established modern practice, the natural circulation of water on the thermo-syphon principle is assisted by a pump to provide what is termed a 'forced-circulation' engine cooling system. This type of system retains the essential components of the simple thermo-syphon system and will be described in detail at a later stage. For the present we will be content merely to list the advantages and disadvantages of the two systems.

10.4.1 Advantages and disadvantages of the thermo-syphon system

(1) A simple arrangement of low cost.
(2) Fairly quick warm-up.
(3) Circulation is proportional only to engine load.
(4) Circulation is not positive.
(5) Large water passages required.
(6) Radiator header tank must have high location relative to jacket.

10.4.2 Advantages and disadvantages of the forced-circulation system

(1) A more complicated and costly arrangement.
(2) Quick warm-up so long as a thermostat is used in the system.
(3) Circulation is proportional both to engine load and speed, the former so long as a thermostat is used in the system.
(4) Circulation is positive.
(5) Smaller water passages are permissible.
(6) Radiator header tank may be lower with respect to the jacket.

10.5 THERMOSTATIC CONTROL OF THE WATER-COOLING SYSTEM

It is worth reflecting that the function of the cooling system is not so much that of keeping the engine cool, but to prevent it from over-heating, in which respect the 'thermostat' plays an important role. The basic purpose of the thermostat is therefore to regulate the flow of coolant through the radiator in accordance with the cooling requirements of the engine. In modern practice, the reasons for using a thermostat may be listed as follows:

(1) Reduce engine warm-up time.
(2) Maintain optimum running temperatures.
(3) Meet the requirements of the car interior heating system.

10.5.1 Operation of the thermostat

The thermostat consists essentially of a poppet control valve, which is opened by a thermal expansion device or element and is closed by a return spring, or return spring action. Movement of the valve is therefore directly related to the temperature of the coolant surrounding its operating element. The thermostat is almost always installed between the coolant outlet from the cylinder head jacket and the inlet to the radiator header tank.

At normal operating temperatures the control valve is open and coolant circulates through both the engine jacket and the radiator, but when cold the valve is closed and circulation through the radiator is prevented, or at least very restricted. The engine is thus allowed to warm up much more quickly, since a smaller quantity of coolant is then being heated by it.

In modern practice, rather than attempting to block completely the coolant flow from the engine during warm-up, either a 'permanent' or a 'variable' radiator bypass arrangement is used with the thermostat and these will be described at a later stage.

10.6 TYPES OF THERMOSTAT

The two types of thermostat that may be used in motor-vehicle cooling systems are as follows:

(1) Aneroid or 'bellows'.
(2) Hydrostatic or 'wax'.

10.6.1 The aneroid thermostat

This type of thermostat consists of a control valve energised by a vapour-filled metal bellows. Control valve movement depends upon the difference between the vapour pressure in the bellows, at any given temperature, and the pressure in the cooling system. The control valve remains closed until the thermostat

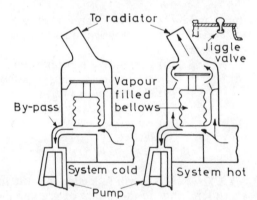

Figure 10.6 Action of the aneroid or bellows thermostat

reaches a predetermined temperature, usually in the range of 75 to 80°C. At this point it begins to open in the direction of coolant flow. It then progressively opens further as the coolant temperature rises, until it is opened fully at around 90 to 95°C. So long as the system is operating, the thermostat continues to control the flow of coolant through the radiator in accordance with engine cooling requirements.

10.6.2 The hydrostatic thermostat

With this type of thermostat, the control valve is energised by an element charged with a wax substance having a high coefficient of thermal expansion. The element consists of a cylindrical metal body containing the wax substance which surrounds a rubber insert, this in turn embracing a central operating thrust rod. As coolant temperature rises, the wax melts and 'compresses' the rubber insert, but since this rubber is constrained to act like an incompressible hydraulic fluid, it

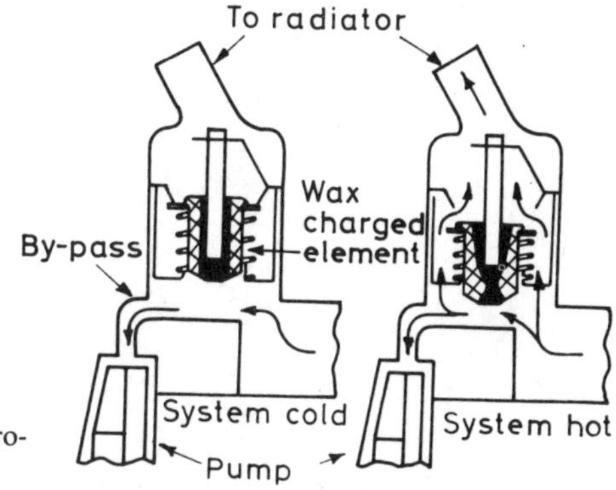

Figure 10.7 Action of the hydrostatic or wax thermostat

displaces the thrust rod. The control valve is thus opened against its separate return spring arrangement. For the modern cooling system, the hydrostatic or wax thermostat has the important advantage that it is relatively insensitive to pressure variations, as discussed next.

10.7 THE RADIATOR PRESSURE CAP

The radiator pressure cap is a combination of filler cap and pressure control valve. It is installed at the highest point in the cooling system and seals against a seating in the filler neck of the radiator. With the engine running and the cap in position,

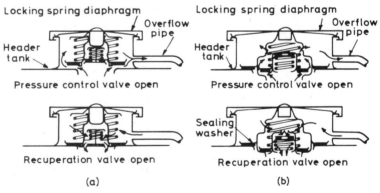

Figure 10.8 Types of radiator pressure cap and their actions: (a) 'open' construction, (b) 'closed' construction

the cooling system is allowed to become pressurised. This occurs automatically because as the temperature increases, the coolant is expanding in a closed system. The reason for using a radiator pressure cap is therefore to maintain the cooling system at a pressure above that of the atmosphere.

10.8 THE PRESSURISED COOLING SYSTEM

The advantages of pressurising the cooling system may be summarised as follows:

(1) It allows the coolant to circulate at a higher temperature

without boiling, so that heat will be transferred more rapidly from the radiator to the air by virtue of the greater temperature difference between them.

(2) It compensates for the reduction in atmospheric pressure when motoring in high-altitude mountainous regions, where the boiling point of the coolant would otherwise be lower with the consequent danger of overheating.

(3) Since the heat transfer from the radiator is directly proportional to the temperature difference between the coolant and the air, a pressurised system allows the radiator to be reduced in size (although it must also be made stronger).

(4) Because the system has to be sealed to pressurise it, instead of simply being vented to the atmosphere, coolant losses through evaporation and surging are minimised.

10.8.1 Operating pressures of radiator cap

The pressure control valve in the radiator cap is usually set to open at a specified value in the range of 28 to 105 kN/m^2. For each 7 kN/m^2 the system pressure is increased above atmospheric, the boiling point of the coolant is raised by about 1.5°C. The recuperation valve in the radiator cap is set to open whenever the system is subject to a depression higher than about 7kN/m^2.

10.9 CONSTRUCTION AND OPERATION OF THE PRESSURE CAP

The radiator pressure cap consists essentially of a spring-loaded plate valve with a rubber facing, which is pre-loaded on to a fixed seating. According to whether an 'open' or a 'closed' type of construction is used, the fixed seating is provided in either the filler neck or the cap itself. If the pressure in the system rises above the limit for which the control valve is rated, the spring loading on it is overcome so that the valve is lifted upwards off its seating and vents the system to atmosphere.

The pressure control valve is in turn fitted with a similar-acting and concentric recuperation valve. This is arranged to open in the opposite direction against light spring loading. Air can therefore be admitted to the radiator to relieve any depression as the coolant temperature falls, otherwise at worst there could be a danger of the header tank collapsing.

CAUTION must be observed in checking the coolant level of a hot engine. The radiator pressure cap should first be part undone to release steam and all pressure, before it is completely removed from the filler neck.

10.10 ANTI-FREEZE SOLUTIONS

The use of an approved anti-freeze solution, all the year round and in all parts of the world, is generally recognised as offering the following advantages:

(1) It protects the cooling system against the damaging effects of frost in cold climates.

(2) The coolant passages are protected against corrosion in both cold and hot climates.

(3) For hot climates there is the incidental advantage that the boiling point of the coolant is raised.

10.10.1 Chemical composition

Of all the various anti-freeze preparations that were once used in motor-vehicle cooling systems, a water-soluble liquid known as 'ethylene glycol' has long been accepted as the most satisfactory, no matter how severe the operating conditions. The adoption of ethylene glycol and water solutions really followed from their earlier use as high-boiling point coolants in piston-type aircraft engines, especially during World War II. Pure ethylene glycol has a boiling point nearly twice that of plain water and this does, of course, explain why only the water evaporates away in an anti-freeze solution. For the same reason, ethylene glycol is sometimes described as a 'permanent' type of anti-freeze material.

So far as the motor-vehicle cooling system is concerned, when ethylene glycol is added to plain water it lowers the freezing point of the solution to well below 0°C. The actual freezing point at which either the water content becomes solid, or the first ice crystals form, is a function of the mixing ratio with water. For example, a 50% by volume concentration of ethylene glycol lowers the freezing point of the coolant to about −37°C. Although higher concentrations of ethylene glycol are irrelevant to motor vehicles, it was in fact found in aircraft engines that an upper limit of about 70% was set by the engine running too hot, because of reduced thermal conductivity of the coolant.

10.10.2 Protecting against corrosion

It was also from experience of the liquid cooling of aircraft engines, which posed problems of metallic corrosion, that the need to include a 'corrosion inhibitor' in the coolant became evident. A similar recommendation was made by the Society of Chemical Industry to the motor industry in 1956 and led in 1959 to the British Standards Institution issuing specifications for three types of inhibited ethylene glycol anti-freeze, known as BS 3150, 3151 and 3152.

The actual chemical composition of these inhibitor systems is of interest only to the chemist, their significance being that BS 3150 was based on a Ministry of Supply aircraft specification and is especially suitable for use in engines of aluminium alloy construction. Specifications BS 3151 and 3152 (the latter being based on American practice) are generally suitable for engines of cast-iron construction. A new British Standard is now in the course of development. The advice of the particular vehicle manufacturer should always, of course, be followed.

10.11 TESTING ANTI-FREEZE SOLUTIONS

Since the degree to which an anti-freeze solution can furnish protection depends upon the mixing ratio of ethylene glycol to water, a hydrometer test is necessary to indicate whether ethylene glycol or water, or both, should be added in the event of the radiator requiring topping-up.

A hydrometer is an instrument for determining the 'relative density' or the number of times a volume of liquid is as heavy as an equal volume of water. It will, no doubt, be more familiar as a convenient method for testing the state of charge of a car battery. In a similar manner, it may be necessary to apply a temperature correction to the reading obtained with an anti-freeze testing hydrometer, otherwise an appreciable error can arise in determining the freezing point of the solution. It is therefore imperative to follow the instructions issued with the hydrometer.

10.12 ENGINE CORE PLUGS

Core plugs are present in the engine cooling jacket for the following reasons:

(1) To blank off the holes left by the jacket cores during casting.
(2) They may be removed for cleaning out corrosive deposits from the jacket.
(3) In the event of the coolant freezing it will expand and force out the core plugs, thereby reducing the risk of cracking the jacket.

10.12.1 Types of core plug

These may be of either the Welch plug, drawn steel cup or, less commonly, the screwed plug variety. The first and second types are respectively expanded and pressed into the machined-to-size core holes.

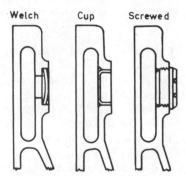

Welch   Cup   Screwed

Figure 10.9 Types of engine core plug

11 Engine ignition system

11.1 THE BATTERY AND COIL IGNITION SYSTEM

To ignite the combustible charge of air and fuel in the cylinder, and thereby initiate the power stroke, some form of 'ignition system' is necessary in the petrol-engined motor vehicle. In fact, it is probably no exaggeration to say that the modern high-speed petrol engine would never have been possible were it not for the remarkable efficiency of its ignition equipment.

11.1.1 Historical background

The earliest source of ignition was, interestingly enough, high-tension electricity, this being used by the Frenchman Etienne Lenoir in 1860 on his famous gas engine, which thus qualifies as the first 'spark ignition' internal combustion engine. However, the then primitive state of development of high-tension electrical equipment did not make for reliability of this type of ignition system. Various other methods of ignition therefore came into use, notably the self-descriptive 'hot-tube' system, which utilised a red-hot platinum tube to initiate combustion in the cylinder.

Quite naturally, the early automobile engineers attempted to adapt for motor vehicle use the existing stationary engine ignition systems, but they soon ran into difficulties. For example, it proved difficult keeping the lamp alight for heating the hot-tube ignition system once the vehicle was in motion! Furthermore, the timing of the ignition control could not be varied, nor could the same time of firing be obtained in each cylinder. Fortunately, by the time the shortcomings of the early ignition systems became evident, the advances that had been made in the design and construction of high-tension electrical equipment were such that more reliable spark-ignition systems became available.

For many years there were two systems of electrical ignition used for motor vehicle engines. These were the self-contained 'magneto' system, which was pioneered by Robert Bosch who founded the famous German firm bearing his name, and the 'battery-and-coil' system developed by Charles F Kettering, the American engineer who also introduced the electric self-starter and was later president of the world-renowned General Motors' Research Laboratories.

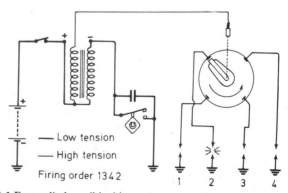

— Low tension
— High tension
Firing order 1 3 4 2

Figure 11.1 Four-cylinder coil ignition system

Although it is now a matter of history that the battery-and-coil ignition system was eventually adopted for all petrol-engined motor cars and commercial vehicles, it may nevertheless be useful to recall the reasons for this, as follows:

(1) The intensity of the spark is reasonably consistent regardless of engine speed and this makes for easy starting.
(2) It has few moving parts, thereby reducing possible sources of unreliability.
(3) There are no heavy components if it is accepted that a battery must in any event be carried for other electric services on the vehicle.

11.1.2 Fundamental requirements of the coil ignition system

An electric spark forms a very convenient means of producing a rapid temperature rise in the engine cylinder, but in practice it requires a very high voltage from the ignition system. For example, with a typical air gap of 0.50 mm at the sparking plug electrodes, the voltage required across them to produce a spark in the engine cylinder is about 10 000 V (10 kV). The qualifying statement 'in the engine cylinder' is made deliberately, because the voltage requirement would be only a few hundred at normal atmospheric pressure. It thus follows that the voltage requirement of the ignition system is raised as the compression ratio of an engine is increased.

Other factors that raise the voltage requirement include surface deterioration of the sparking plug electrodes, the burning of weak mixtures and cold starting of the engine. The latter is partly explained by the lower temperature of the cylinder gases surrounding the plug electrodes because, unlike a metallic conductor, the electrical resistance of a gas decreases when it is heated.

Apart from actually producing the spark, the next requirement is the not inconsiderable one of the number of sparks necessary in a given interval of time. Take, for example, a four-cylinder (four-stroke) engine where there are two power strokes for every complete revolution of the crankshaft. This means that the ignition system must likewise fire two sparking plugs for every crankshaft revolution. At, say, 4500 engine revolutions per minute (75 per second), the ignition system must therefore produce 150 high-voltage sparks every second. This figure would, of course, be doubled in the case of an eight-cylinder engine running at the same speed.

Furthermore, it is an important requirement of the ignition system that the high-voltage spark must be timed to occur in each cylinder at an optimum point in the engine operating cycle.

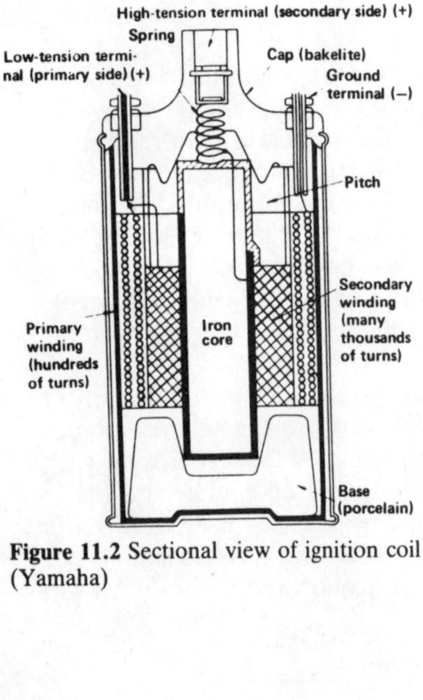

Figure 11.2 Sectional view of ignition coil (Yamaha)

11.1.3 Basic action of the coil ignition system

At this stage our attention will be confined to the basic action of the battery-and-coil ignition system, without reviewing the underlying theory of electromagnetic induction upon which the system relies for its operation.

A general understanding of the coil-ignition system may be gained by referring to the typical circuit diagram for a four-cylinder engine application. The first point to notice here is that the ignition system comprises two interacting circuits, these being known as 'low-tension primary' and 'high-tension secondary'. Their meeting place is in the ignition coil itself, where their respective windings of a few hundred and many thousand turns of wire act in the manner of a transformer and provide the necessary step-up in the battery and charging system voltage to fire the sparking plugs.

Unlike an ordinary transformer, however, which utilises an alternating current, the ignition coil is supplied with a fluctuating direct current from the battery and charging system. This is because as the engine rotates, the 'contact breaker' in the ignition 'distributor' switches on and off the current flowing through the primary circuit winding in the coil. The important effect of this sudden switching action is to produce a corresponding reaction in the secondary circuit winding in the coil, such that if a relatively small voltage is applied to the primary winding a voltage of many thousand times this value will occur across the secondary winding.

A further point to notice in the circuit diagram is that the secondary winding of the ignition coil is connected to the central electrode of the distributor 'cap' and thence to what is termed the 'rotor arm' which, like the contact breaker, is driven at half engine speed. At the same instant as the high-tension voltage occurs across the secondary circuit, the conducting part of the rotor arm also becomes aligned with one of the ring of four internal electrodes in the distributor cap. This immediately results in the high-tension current jumping the small air gap to the electrode and then travelling through an insulated ignition 'lead' to fire the appropriate sparking plug.

11.2 COIL IGNITION EQUIPMENT

The conventional battery-and-coil ignition system comprises the following units:

(1) *Battery and charging system:* This is the source of supply of electrical energy for the low-tension primary circuit of the ignition system.

(2) *Ignition switch:* The purpose of this is to connect and disconnect the ignition system from the battery and charging system so that the engine can be started and stopped by the driver.

(3) *Ignition coil:* The purpose of this is to transform voltage from a low-tension source, that is the battery and charging system, into a high-tension voltage sufficient to promote an electrical discharge across a fixed air gap at the sparking plug.

(4) *Ballast resistor:* This may be added to the ignition coil for the purpose of improving the life of the contact breaker points, since except under starting conditions the resistor reduces the current flowing through the ignition coil primary circuit.

(5) *Contact breaker:* This is a cam-actuated interrupter switch contained in the ignition distributor, its function being to open and close the ignition coil primary circuit.

(6) *Condenser:* This is connected across the contact breaker points and provides temporary storage for electric energy as they open, thereby minimising arcing that would otherwise shorten their life.

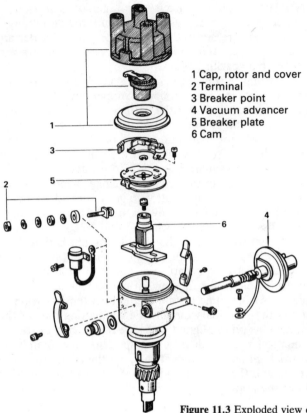

1 Cap, rotor and cover
2 Terminal
3 Breaker point
4 Vacuum advancer
5 Breaker plate
6 Cam

Figure 11.3 Exploded view of ignition distributor (Toyota)

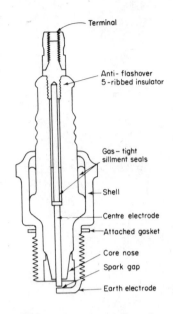

Figure 11.4 Sectioned view of sparking plug (Champion)

(7) *Rotor arm and distributor cap:* In combination these provide a rotary switch that receives the high-tension current from the ignition coil each time the primary circuit is interrupted, and then distributes it to the appropriate cylinder sparking plug.

(8) *Ignition distributor:* Apart from performing the functions described under (5) and (7), it must also incorporate a mechanism for automatically varying the ignition timing in accordance with engine operating requirements.

(9) *Ignition leads:* These are heavily insulated cables conveying the high-tension current from the coil to the distributor and thence to the sparking plugs.

(10) *Sparking plugs:* Their purpose is to conduct the high-tension current from the ignition system into the combustion chambers of the engine cylinders, wherein they promote an electric discharge across a fixed gap between their electrodes to ignite the combustible charge.